Getting Started with CNC

Edward Ford

Getting Started with CNC

by Edward Ford

Copyright © 2016 Maker Media. All rights reserved.

Printed in the United States of America.

Published by Maker Media, Inc., 1160 Battery Street East, Suite 125, San Francisco, CA 94111.

Maker Media books may be purchased for educational, business, or sales promotional use. Online editions are also available for most titles (*http://safaribooksonline.com*). For more information, contact O'Reilly Media's institutional sales department: 800-998-9938 or *corporate@oreilly.com*.

Editor: Roger Stewart
Technical Reviewer: Jonathan Ward
Production Editor: Nicholas Adams
Copyeditor: Jasmine Kwityn
Proofreader: Sharon Wilkey
Indexer: Judith McConville
Interior Designer: David Futato
Cover Designer: Julie Cohen
Illustrator: Rebecca Demarest

August 2016:   First Edition

Revision History for the First Edition

2016-08-04:   First Release

See *http://oreilly.com/catalog/errata.csp?isbn=9781457183362* for release details.

Make:, Maker Shed, and Maker Faire are registered trademarks of Maker Media, Inc. The Maker Media logo is a trademark of Maker Media, Inc. *Getting Started with CNC* and related trade dress are trademarks of Maker Media, Inc.

Many of the designations used by manufacturers and sellers to distinguish their products are claimed as trademarks. Where those designations appear in this book, and Maker Media, Inc. was aware of a trademark claim, the designations have been printed in caps or initial caps.

While every precaution has been taken in the preparation of this book, the publisher and authors assume no responsibility for errors or omissions, or for damages resulting from the use of the information contained herein.

978-1-457-18336-2

[LSI]

WITHDRAWN

Contents

Preface........ ix

1/What Is CNC?........ 1
   Digital Fabrication........ 2
   Why Computer Controlled?........ 2
      Accuracy........ 2
      Complexity........ 3
      Simulation........ 3
      Safety........ 4
   How Do Computer-Controlled Machines Work?........ 6
      Cartesian Coordinate System........ 6
      X, Y, and Z for CNC........ 7
   What Can I Make?........ 9
      Toys and Games........ 9
      Signs and Carvings........ 12
      Vehicles, Furniture, and Houses........ 14
      Molds and Casts........ 16
      Metal Creations and Inlays........ 17
      Circuit Boards........ 20

2/Mechanical Overview........ 23
   Gantry........ 24
   Carriage........ 24
   Spindle........ 24
      Spindles Versus Routers........ 26
   Table........ 28
   Mechanisms for Securing Materials........ 28
      Step Clamps........ 29
      T-slots........ 31
      Threaded Inserts........ 32
      Screws........ 32
      Tape........ 33
      Vacuum Table........ 33
      Vises........ 34

Routers Versus Mills. 35
Machine Configurations. 36

3/End Mills and Cutting. 39
   End Mills. 39
      Drill Bits Versus End Mills. 40
      Common Tool Geometries. 42
      Tip Shapes. 45
      End Mill Anatomy. 45
      End Mill Materials. 46
      Coatings. 47
   Cutting. 47
      Ramping. 48
      Climb Versus Conventional Cuts. 48
      Speeds and Feeds. 49
      Chipload. 50
      Types of Tool Holding. 51

4/CAD: Draw or Model Something. 55
   2D Raster Images. 56
   2D Vector Graphics. 57
      Vector Editing Software. 58
   2D Drawings Versus 3D Models. 60
   3D Models. 62
   More Software to Try. 63
      V-Carving Text. 63
      Image to G-code. 63
      Single-Line Drawing. 64
      Halftone Images. 65

5/CAM: Make Toolpaths. 67
   2D/2.5D Toolpaths. 68
   3D Toolpaths. 69
   2D/2.5D CAM Operations. 69
   2D/2.5D Toolpath Parameters. 71
   Overcuts. 73
      Dog Bones. 75
      T-Bones. 76
   Minimum Feature Size. 77
   Basic 3D CAM Operations. 79
      Parallel Finishing. 79
      Contour Finishing. 80

6/CAD/CAM Project: No Machine Necessary!. 81
Inkscape. 81
MakerCAM. 81
Webgcode. 82
CAMotics. 83
Wooden Racer Project. 83
    Project Materials and Dimensions. 84
Step 1: Create the Digital Design. 85
    Body. 85
    Wheels. 86
Step 2: Configure MakerCAM. 87
Step 3: Import and Center Racer SVG File. 87
Step 4: Create Wheel Toolpaths. 88
    Reduce Wheel Thickness by Half. 89
    Screw Head Countersink. 89
    Screw Hole. 90
Step 5: Create Body Toolpaths. 91
    Body Window. 91
    Wheel Holes. 91
    Body Perimeter. 92
Step 6: Calculate Toolpaths. 92
Step 7: Export G-code. 92
Step 8: Vizualize Toolpaths. 92

7/Creating Motion: Electromechanical Overview. 95
Mechanical Motion. 95
    Linear Motion. 95
    Slop. 96
    Linear Guide Types. 96
Power Transmission. 97
    Lead Screws and Lead Nuts. 98
    Belt Drives. 99
    Rack and Pinion. 99
Backlash. 100
Motors and Electronic Components. 101
Stepper Motors. 102
Motion and Machine Control. 105
    Mach3. 106
    LinuxCNC. 107
    Grbl. 107
    Industrial Cases. 108

    Other Choices. 108
    Parallel Ports. 109

8/G-Code: Speaking CNC. 111
    Drawing a Square: Instructions for Humans. 112
    Square-Drawing Instructions for Machines. 113
    G-code Square Breakdown. 114
        Step 1: Put Pen to Paper (G20 F20 X0 Y0 Z0). 114
        Step 2: Move the Pen 1 Inch Toward the Top (G1 Y1). 115
        Step 3: Move the Pen 1 Inch Right (G1 X1). 115
        Step 4: Move the Pen 1 Inch Toward bottom (G1 Y0). 116
        Step 5: Move the Pen 1 Inch Left (G1 X0). 117
        Step 6: Lift the Pen 1 Inch from Paper (G1 Z1). 117
    G-code Rules. 118
    Feeds, Speeds, and Tools. 118
    Diving Further into G-Codes. 119
        G0 (Rapid Motion). 119
        G1 (Controlled Motion). 120
        G2 (Clockwise Motion). 121
        G3 (Counterclockwise Motion). 122
        G17/G18/G19 (Working Planes). 123
        G20/21 (Inches or Millimeters). 123
        G28 and G28.1 (Referencing Home). 123
        G90 (Absolute Mode). 124
        G91 (Incremental Mode). 124
    M-Codes. 124

9/Practical Machining Tips. 127
    CAM File Orientation Versus Actual Machine Setup. 127
    Setting Machine Zero. 128
        Zeroing the X- and Y-Axes. 128
        Zeroing the Z-Axis. 130
    Homing. 132
        Practical Homing. 132
    Tool Changes. 133
    Machined Material Hold-Down Tips. 134

10/Conclusion. 137

A/Resources. 139

Index. 143

Preface

Fifteen years ago, I watched a CNC machine run for the first time. It was mesmerizing. After observing it for a few weeks, I decided that I would build a small one—I had to. *I was in love.*

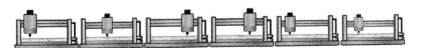

As I began to gather information, I found myself *slowly* peeling back the layers, but I had no idea what I was getting into—you don't know what you don't know. Around every corner, I'd encounter a new term, picture, or video documenting something completely foreign. I'd pursue these leads, but as I began to grasp one concept, it would spawn another series of questions.

After two *long* years of research, I decided that I had finally gathered enough information to begin creating my own CNC machine. The first prototype wasn't accessible, affordable, or able to be replicated, but building it was great practice for subsequent attempts (of which there were many).

During the summer of 2009, I began designing and building a small CNC I dubbed *Shapeoko*, named for the Shapeways 3D printing and Ponoko laser cutting services I used to create custom parts for the initial prototypes.

Two years later, I Kickstarted Project Shapeoko so I could share what I had learned through a long process of trial and error as I prototyped a $300 dead-simple CNC machine (Figure P-1). I wanted to make its creation repeatable and the plans freely distributable for everyone to enjoy. An active community sprang up around the project, asking and answering questions both in the forum (*http://www.shapeoko.com/forum*) and the wiki (*http://www.shapeoko.com/wiki*).

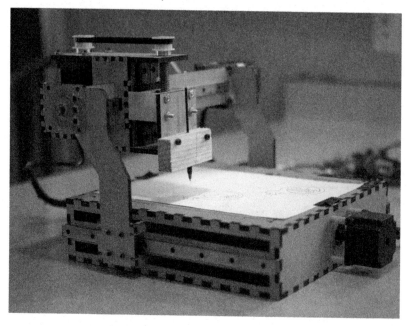

Figure P-1. *Project Shapeoko prototype that launched the open hardware Kickstarter in 2011*

Why I Wrote This Book

As with Project Shapeoko, I wrote this book to flatten the CNC learning curve, "kickstarting" the reader's own journey into CNC. But most of all, I want to share all of the things that I wished someone would have told me about CNC when I first started.

Who This Book Is For

This book provides a basic overview of affordable, hobbyist-level computer-controlled routers. It's helpful to have a general understanding of how to draw on a computer, especially using 2D vector graphics, but no prior CNC knowledge or experience is required—*you don't even need CNC access to complete the exercises in this book.*

If you're a Maker who wants to learn what CNC machines are, how they work, and how to use them, but you've been intimidated by the terminology or had no idea what questions to ask, then this book is for you!

How to Use This Book

Each chapter of *Getting Started with CNC* deals with a discrete topic, from the basics of computer-aided design to a deep dive into the meaning of individual G-codes.

Beginners should approach this book as a general introduction to desktop CNC routing and read it cover to cover. Those with CNC experience will still find it useful as a reference book, particularly Chapter 8.

Although I've covered the CNC basics to get you started, this book is not all-inclusive—it's just the first step into the world of CNC. After you've finished reading this book, consult Appendix A, which lists additional resources that will help as you continue exploring and growing your skill set.

Conventions Used in This Book

The following typographical conventions are used in this book:

Italic
    Indicates new terms, URLs, email addresses, filenames, and file extensions.

`Constant width`
    Used for program listings, as well as within paragraphs to refer to program elements such as variable or function names, databases, data types, environment variables, statements, and keywords.

`Constant width bold`
    Shows commands or other text that should be typed literally by the user.

This element signifies a tip, suggestion, or general note.

 This element indicates a warning or caution.

Using Code Examples

This book is here to help you get your job done. In general, you may use the code in this book in your programs and documentation. You do not need to contact us for permission unless you're reproducing a significant portion of the code. For example, writing a program that uses several chunks of code from this book does not require permission. Selling or distributing a CD-ROM of examples from Make: books does require permission. Answering a question by citing this book and quoting example code does not require permission. Incorporating a significant amount of example code from this book into your product's documentation does require permission.

We appreciate, but do not require, attribution. An attribution usually includes the title, author, publisher, and ISBN. For example: "Getting Started with CNC by Edward Ford (Make:). Copyright 2016, 978-1-457-18336-2."

If you feel your use of code examples falls outside fair use or the permission given here, feel free to contact us at *bookpermissions@makermedia.com*.

Safari® Books Online

 Safari Books Online is an on-demand digital library that delivers expert content in both book and video form from the world's leading authors in technology and business.

Technology professionals, software developers, web designers, and business and creative professionals use Safari Books Online as their primary resource for research, problem solving, learning, and certification training.

Safari Books Online offers a range of plans and pricing for enterprise, government, education, and individuals.

Members have access to thousands of books, training videos, and prepublication manuscripts in one fully searchable database from publishers like O'Reilly Media, Prentice Hall Professional, Addison-Wesley Professional, Microsoft Press, Sams, Que, Peachpit Press, Focal Press, Cisco Press, John Wiley & Sons, Syngress, Morgan Kaufmann, IBM Redbooks, Packt, Adobe Press, FT Press, Apress, Manning, New Riders, McGraw-Hill, Jones & Bartlett, Course Technology, and hundreds more. For more information about Safari Books Online, please visit us online.

How to Contact Us

Please address comments and questions concerning this book to the publisher:

> Make:
> 1160 Battery Street East, Suite 125
> San Francisco, CA 94111
> 877-306-6253 (in the United States or Canada)
> 707-639-1355 (international or local)

Make: unites, inspires, informs, and entertains a growing community of resourceful people who undertake amazing projects in their backyards, basements, and garages. Make: celebrates your right to tweak, hack, and bend any technology to your will. The Make: audience continues to be a growing culture and community that believes in bettering ourselves, our environment, our educational system—our entire world. This is much more than an audience, it's a worldwide movement that Make: is leading—we call it the Maker Movement.

For more information about Make:, visit us online:

Make: magazine (*http://makezine.com/magazine/*)
Maker Faire (*http://makerfaire.com*)
Makezine.com (*http://makezine.com*)
Maker Shed (*http://makershed.com/*)

We have a web page for this book, where we list errata, examples, and any additional information. You can access this page at *http://bit.ly/getting-started-with-cnc*.

To comment or ask technical questions about this book, send email to *bookquestions@oreilly.com*.

Acknowledgments

After 15 years of working with CNC machines, I still find myself learning something new almost every day.

I'd like to thank all of the people who helped me understand CNC, including the following:

- My dad, Gerald Ford, has worked in industrial manufacturing my entire life. When I was a child, I would occasionally go with him to work on a weekend morning. I have fond memories of watching machines being operated, listening to all the sounds that a shop floor makes. And the smell. My father taught me the fundamentals of manufacturing and the fundamentals of being a good human being.
- My grandfather, Stuart Keller, passed away while I was writing this book. He was a former radar mechanic for the Navy, a physics professor, and a ham radio enthusiast. Grandpa taught me about electronics, showed me how to solder, and introduced me to programming in BASIC.
- Patrick Reaver, my buddy and confidant. Patrick taught me how to use Autodesk Inventor and was always quick to constructively point out design flaws.
- Bart Dring, the creator of MakerSlide, and a good friend. Bart's design style and his belief in open hardware and the Maker movement has always inspired me.

- My partners at Carbide 3D: Rob Grzesek, Jorge Sanchez, and Apollo Crowe. Thank you for the technical references, the massive editing (Jorge!), and the great project pictures (Apollo).
- The Gentlemen of B.C.: Anthony, Brandon, Lorenzo, Nate, Pat, Patrick, and Tommy. Thanks for all of the support.
- The Original Shapeoko Crew: Brandon Fischer, Tim Foreman, Will Adams, and Winston Moy. You guys are all great, and have helped Shapeoko in so many ways since this project started five years ago. Special thanks to Will especially for helping proofread, critique, and review this book.

Last, but not least, I'd like to thank my wife, Laura. I could write an entire book on how much I love her. She and our two children are who inspire me to be the best person I can be. Every day.

1/What Is CNC?

A *CNC* is a machine whose movements are controlled by a computer through a process known as *Computer Numerical Control*. Some examples of common CNC machines are routers, milling machines, 3D printers, vinyl cutters, laser cutters, pick-and-place machines, and many others.

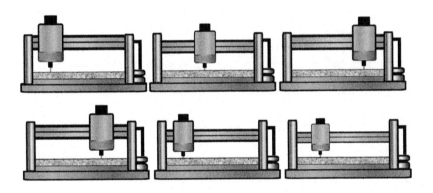

Figure 1-1. *CNC router in action*

Although we'll be talking mostly about subtractive CNC machines that remove material with sharp, spinning tools (as illustrated in Figure 1-1), the principles of CNC apply to any type of CNC machine you may encounter.

In the Maker lexicon, it's common to refer to a computer-controlled router or mill as a *CNC*, while explicitly naming other machine types. For example, "I have a 3D printer, a laser cutter, and a CNC."

Digital Fabrication

Because CNCs are by definition computer controlled, they need to ingest computer-created instructions that tell them *exactly* how to move. *Digital fabrication* is the process of designing an object in software (CAD, or computer-aided design), defining the way the machine will move to create that object (CAM, or computer-aided manfacturing), and then physically running the machine to turn those previously created codes into tangible, real-world items.

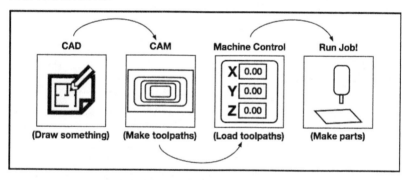

Figure 1-2. *Typical digital fabrication workflow*

Why Computer Controlled?

There are several advantages to controlling a machine with a computer over manual operation: accuracy of the cuts, complexity of the parts, ability to simulate the job in software, and overall safety.

Accuracy

Computer-controlled machines move *exactly* as instructed. If you need the machine to move exactly 1.2 inches between slots or drill a hole that's precisely 1.75 inches deep, no problem! Numerically controlled machines can do these tasks far more reliably and repeatably than we can do manually.

Complexity

Many intricate designs are very difficult to achieve with manual tools—especially if you're not a master craftsperson. CNCs can make intricate cuts and carve advanced three-dimensional shapes that may be all but impossible to do with hand-controlled tools.

Nicholas Fortosis created the beautiful window inlay shown in Figure 1-3 with his Shapeoko CNC machine.

Figure 1-3. *Complex window screen design*

Simulation

CAM software allows you to preview (as illustrated in Figure 1-4), or *simulate*, how the machine will move in the physical world as it cuts your digital file. It's also common practice to *dry run* your files by sending the instructions to the CNC machine without a bit in place. Then you can watch the machine move, ensuring it stays within the cutting area limits and verifying that there are no errors to prevent it from finishing.

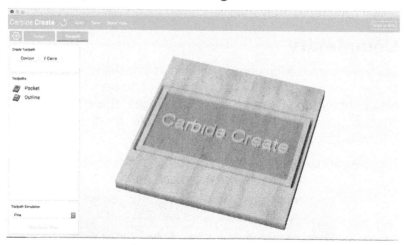

Figure 1-4. Toolpath simulation in a CAM program

 With CNC simulation built into both free and commercial CAM packages, it's not only trivial to preview your job pre-machining, but negligent not to!

Safety

Any automated machine or power tool (manual or CNC) with spinning blades has inherent dangers. If safety precautions are not followed carefully, serious injury can result.

Although CNC movement can be simulated, and the ability to keep a safe distance from the machine while in operation and to use emergency stop buttons, or *kill switches*, help make computer-controlled machines safer, there are still inherent dangers. Knowing where the E-stops (shown in Figure 1-5) are located is always a good idea.

Figure 1-5. *Emergency stop button on a Tormach PCNC 1100 personal CNC mill*

Safety First!

As any shop teacher or machinist can tell you, there are dangers associated with all tools, automated or manual. Here are some safety rules you should *always* follow to avoid injury:

- Wear safety glasses—bits can snap, debris can break loose, and excessive dust can irritate the eyes.
- No loose-fitting garments—sleeves or scarves can easily get snagged on or even pulled into a machine.
- No jewelry, especially rings and necklaces.
- Tie back long hair.
- Do not use visually worn or broken tooling.
- Stay close to the emergency stop button (Figure 1-5).
- Wear hearing protection.
- Have a first aid kit handy.

What Is CNC? 5

How Do Computer-Controlled Machines Work?

In order for a CNC machine to make precise cuts, it needs to know where and how quickly to move in three-dimensional space. These movement instructions, or *toolpaths*, are called *G-code* (we'll discuss G-code in more depth in Chapter 8). G-code is created in a CAM program and then typically sent to the machine over a physical connection (usually a USB cable).

Cartesian Coordinate System

You probably learned about the *Cartesian coordinate system* in high school geometry class. It defines points, or *coordinates*, numerically in two- or three-dimensional space. Here is a brief review of a few key terms:

Axes
   Axes are number lines that intersect at right angles. The horizontal line, or x-axis, runs perpendicular to the y-axis. The numeric position where the two lines intersect is a coordinate (see Figure 1-6). Coordinates can be composed of both negative and positive numbers, depending on the quadrant where they are located.

Two-dimensional space (x- and y-axes)
   Coordinates in two-dimensional space are defined by an ordered pair that indicates the perpendicular intersection of horizontal (x-axis) and vertical (y-axis).

Three-dimensional space (adding the z-axis)
   When working in three-dimensional space, a third axis must be added. This additional dimension—called the z-axis—represents depth.

Without depth, CNC machining as we know it would not be possible. At the most basic level, the z-axis allows the machine to raise and lower the bit.

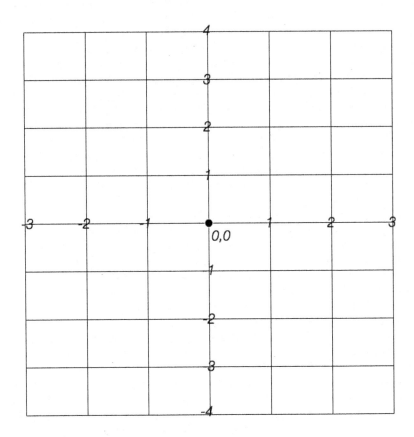

Figure 1-6. *Two-dimensional Cartesian diagram*

X, Y, and Z for CNC

Most desktop CNC machines have three axes, as shown in Figure 1-7.

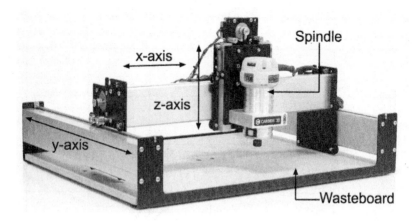

Figure 1-7. *A CNC router with x-, y-, and z-axes labeled*

When facing a CNC machine from the front, the x-, y-, and z-axes are typically arranged and move in the following fashion:

z-axis
    As shown in Figure 1-7, the *carriage* that holds the *spindle* or *router* is attached to the z-axis. It moves up above the machine surface when the z is a positive number and down when z is negative.

x-axis
    The x-axis runs horizontally across the front of the machine, moving left and right. It also supports the z-axis, which moves up and down while sliding left to right on the x-axis.

    When the x-axis moves horizontally to the right, the *x* coordinate becomes a steadily larger number. The opposite is true for left movements, which progressively shrink the *x* coordinate until the end of the axis is reached.

y-axis
    In CNC routers, the y-axis is typically fixed, as seen in Figure 1-7. Just like the y-axis in a two-dimensional graph, it runs perpendicular to the x-axis, which slides back and forth between the front and back of the machine. As the y-axis moves away from you, it's moving in the positive direction. When moving toward you, the *y* coordinate is becoming

steadily smaller, until it reaches the front (or negative limit) of the machine.

What Can I Make?

CNC routers are versatile. They're capable of cutting and engraving many different types of materials: hard and soft woods, plastics, and soft metals such as aluminum and brass.

DIY machine sizes and styles vary widely, from small desktop models to large shop units capable of cutting sizable items from sturdy materials. The combination of these multimaterial capabilities with digital design is quite powerful, enabling you to build nearly anything imaginable—from wooden toys to an entire house!

Toys and Games

A quintessential part of childhood, wooden toys (such as the car shown in Figure 1-8) have been around almost as long as children. Traditionally, these beautiful toys were made with a scrollsaw and other non-CNC tools, but with a CNC machine, they can be customized and designed with a more organic shape.

Figure 1-8. *Pinewood derby car made by @apolloness on his Nomad 883*

What Is CNC? 9

Inspired by a Pegs and Jokers game at a friend's house, Stephen and Cara Bell decided to make their own board, shown in Figure 1-9 (you can read more about their project at the Shapeoko website (*http://bit.ly/2a6DAc2*)). They designed the board in Illustrator and machined from 3/8-inch poplar (purchased at Home Depot) using a 1/4-inch bit and MakerCAM (*http://Makercam.com*). After sanding and clear-coating, it was time to play!

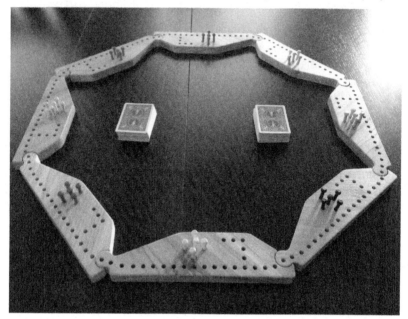

Figure 1-9. *Stephen and Cara Bell's Pegs and Jokers board game*

Darren Lafreniere runs TOhBaby.com (*http://tohbaby.com/*), a site that features CNC'ed baby teethers, rattles, blocks, and puzzles. Darren used a vise to align four blocks at once to cut patterns into the wooden puzzle blocks of various woods shown in Figure 1-10 (you can read more about his project at the Carbide 3D website (*http://bit.ly/29PvAcD*)).

Figure 1-10. *Wooden puzzle blocks*

Tim Foreman created a crossbow pistol that shoots mini-marshmallows, shown in Figure 1-11. After posting a build log in the Shapeoko forums (*http://bit.ly/29QmnBI*), he wrote up an extensive step-by-step tutorial (*http://bit.ly/29VJnIE*).

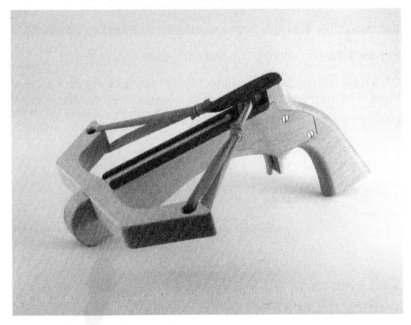

Figure 1-11. *Tim Foreman's mini-marshmallow crossbow pistol*

Signs and Carvings

Professional woodworkers commonly use CNCs paired with V-carving software to speed up manual tasks, creating beautiful signage by importing existing images or creating custom designs (see Figure 1-12). Popular sign-making materials include hardwoods, RenShape, and two-color high-density polyethylene (HDPE) plastics.

Figure 1-12. *US Navy sign by Shapeoko user @markspens*

CNCs are also used to create highly detailed engraved creations, like that shown in Dan Spangler's carbonite foam relief (see Figure 1-13).

Figure 1-13. *Han Solo in carbonite foam relief by Dan Spangler*

Vehicles, Furniture, and Houses

Large-format CNCs capable of cutting standard 4-inch × 8-inch sheets of plywood are now widely available at MakerSpaces, schools, and shared shops all around the world, making the creation (and replication) of large-scale objects easier and more affordable than ever before.

With a large enough cutting area, you can make just about anything imaginable—even vehicles, furniture, and houses!

The gas-powered PlyFly Go Kart is shown in Figure 1-14. The PlyFly kit is fabricated and sold by Flatworks LLC (*http://theflatworks.com/*).

Figure 1-14. *Assembled PlyFly Go Kart*

As the "create globally, fabricate locally" movement grows, everything from simple bedside tables to entire furniture sets are being designed and freely distributed for CNC machining by companies such as AtFAB (*http://atfab.co*) (see Figure 1-15).

Figure 1-15. *The AtFAB downloadable furniture collection*

Online communities like 100kGarages (*http://www.100kgarages.com/*), Opendesk (*https://www.opendesk.cc/*), and FabHub (*https://www.fabhub.io/*) now connect designers, fabricators, and consumers by making design files available DIY or by finding a local fabricator to produce the furniture design.

Check out Make:'s "Where to Get Digital Fabrication Tool Access" page (*http://bit.ly/29QrzYJ*) to locate machines near you!

The beauty of CNC is in precision and predictability. Once a project's digital design is complete and tested, it can be infinitely replicated on any capable CNC. Even large structures, such as the 14-foot × 16-foot CNC'ed Makerspace Shed (*http://bit.ly/29VJAWe*) (pictured in Figure 1-16) can be created quickly from cheap materials like oriented strand board (OSB).

Figure 1-16. *CNC'ed Makerspace Shed*

Molds and Casts

With a micro-resolution CNC, you can make the tiniest details accurately and repeatably. Michal Zalewski milled medium-density modeling board plastic (Figure 1-17) to create molds for resin casting miniscule robotics parts, as shown in Figure 1-18. You can read more about his project at his website (*http://lcamtuf.coredump.cx/gcnc/*).

Figure 1-17. *Michal Zalewski's machined wax mold for resin casting*

Commonly referred to as *resin casting*, this process entails creating the shapes you wish to cast from your CAD model, machining it, and then casting with silicone.

Once the silicone has cured, your mold is complete. From there, you only need to select a resin, add a little color (if that's your thing), and pour the resin into the mold. In as little as a few hours, you'll have finished, high-quality durable parts.

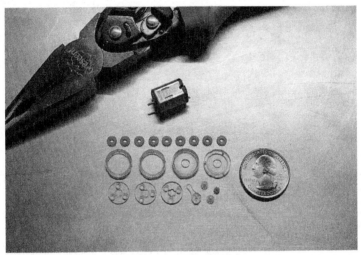

Figure 1-18. *Cast parts by Michal Zalewski*

Metal Creations and Inlays

Desktop CNC machines are capable of cutting a variety of materials, from foam to metals such as aluminum (shown in Figure 1-19). Joe Ternus made a one-of-a-kind ring box, shown in Figure 1-20, to hold a very special engagement ring (she said yes!). Joe's ring box is a perfect example of what can be done with a CNC machine with enough imagination and prototyping patience.

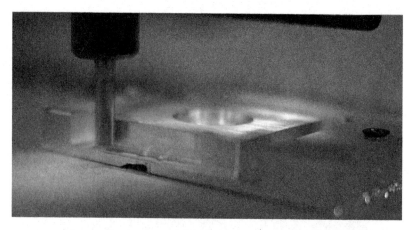

Figure 1-19. *Routing an aluminum part*

Figure 1-20. *CNC milled ring box by Shapeoko user Joe Ternus*

Jewelry designers can also benefit from computer-controlled machines. As reported on the Carbide 3D forum (*http://bit.ly/29VJSfJ*), the purchase of a small, accurate CNC mill enabled the formerly part-time proprietor of Camillette Jewelry (*http://camillette.com/random-collection/*) to quit her day job (Figure 1-21 shows one of Camillette's finished pieces). Designs that previously entailed tedious manual cutting are now easily replicatable and make a small business scalable.

Figure 1-21. *Camillette Jewelry CNC'ed brass ring*

In addition, the fine tolerances needed to precisely fit one material into another to create intricate inlays is *de rigueur* with machining (Figure 1-22).

Figure 1-22. *Inlayed bottle opener*

Circuit Boards

With a CNC at your disposal, you can skip messy toner transfers or chemical etchants. You can also skip sending out your *printed circuit board* (PCB) designs to an online service, where you'll wait weeks to find your first mistake.

CNC machines allow you to convert your PCB files into CNC G-code, and then create your design right there on your machine the same day you design it.

Carbide 3D user Marc Liyanage created a beautiful TQFP-44 breakout board with his Nomad CNC machine (see Figure 1-23).

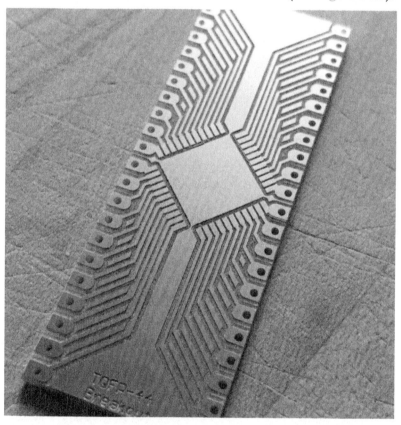

Figure 1-23. *TQFP-44 breakout board by Marc Liyanage*

This book focuses mainly on three-axis desktop CNC routers that cost less than $5,000. However, the world of computer-controlled machinery encompasses so much more. There's a CNC machine behind nearly every manufacturing application imaginable, from six-axis robotic arms that spray a perfect paint finish on an exotic car, to a 6,000-watt laser cutting through 1-inch-thick steel.

The machines we will look at are intended for a wide range of uses—some are more hobby driven and not meant to run in a production environment, while others are more robust and capable of running longer duty cycles. Regardless of the machine's quality or robustness, the operation (perhaps not the results) are the same across the board.

In the chapters that follow, we'll look at all of the major systems associated with the machines, along with the software and workflow required to go from part to idea.

2/Mechanical Overview

To the untrained eye, all CNC machines look different and appear to work from a different set of components.

After becoming familiar with the underlying mechanics of CNC machines, it's easy to spot a pattern and recognize that almost all CNC machines are designed the same way.

CNC machines can be divided into a few key parts, as shown in Figure 2-1. CNCs come in many different styles, sizes, and shapes, but there are basic machine features regardless of layout.

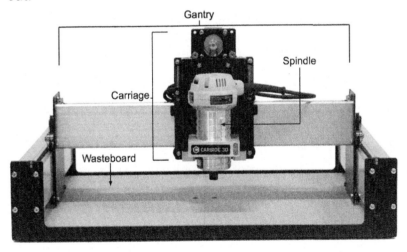

Figure 2-1. CNC router diagram with labeled key parts

Gantry

The *gantry* spans the x-axis (left to right) and travels forward and backward along the side y-axis rails. Attached to the gantry are the carriage and spindle, which make up the z-axis.

 The *available travel* (the distance the carriage can move before hitting the side, or *limit*, of the machine) dictates the maximum part width that can be cut.

Carriage

The *carriage* is the assembly on the machine that carries the z-axis rails and spindle. It travels left to right on the gantry along the x-axis. A motor raises and lowers the spindle up and down along the z-axis.

Spindle

The *spindle* is attached to the carriage and holds a cutting tool, called an *end mill*, which bites into the material being machined as the spindle rotates it clockwise. End mills are described in detail in Chapter 3.

Regardless of how rigid your machine may be, how quickly it can move, or how accurately it can position itself, if your spindle is junk, parts will never turn out as nicely as you want them to, because of increased runout.

The spindle is made up of several key components: the body, neck (or shank), shank end, collet, and collet nut. Here's a brief overview of each of these parts (see Figure 2-2):

Body
    The body of the spindle is where the motor, bearings, and rotational shaft are located.

In some cases, the motor is separate from the body, but this is not typical.

Neck/shank
The shank protrudes below the body of the spindle. The longer the shank's distance, the more room you have around the spindle to avoid obstacles such as clamps and hardware.

This is because the neck/shank of the spindle or router typically has a much smaller diameter than the body of the spindle or router. Because the diameter is smaller, you can get the neck lower without needing as much room for clearance.

Collet
The collet is the collar that holds the end mill in place. They come in various sizes and configurations, and must be matched with the spindle type (as explained in "Types of Tool Holding" on page 51).

Collet nut
Threads onto the end of the shank while squeezing the collet around the tool.

Shank end
Threaded portion at the end of the shank for securing the collet nut that keeps the collet in place.

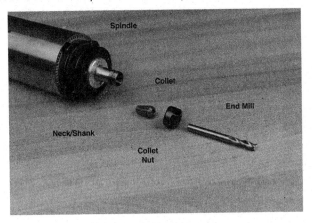

Figure 2-2. *1.5kw spindle with labeled components*

Spindles Versus Routers

The terms *router* and *spindle* are often used interchangeably, as both are spinning the cutting bit, but technically they refer to two different types of tools (see Figure 2-3).

Figure 2-3. *1.5kw air-cooled spindle (left) and trim router (right)*

Routers are an easy and affordable solution; they are widely available and easy to hook up. Most routers can be plugged directly into a traditional wall outlet, and turned on with the built-in on/off switch.

 Router speed control can be accomplished through a traditional router speed controller, but in most cases the router loses torque as the speed is reduced.

Spindles are more expensive, and require a *variable frequency drive* (VFD) for control. Spindles can maintain full torque at much lower RPMs compared to traditional routers. They also have precision bearings, which mean less *runout* (or tool rotation inaccuracy). Spindle cuts are smoother, quieter, and more accurate than those made with a router.

In addition, spindles are much quieter in operation than routers because of how they are cooled. Spindles are generally water cooled, whereas routers are air cooled, requiring a fan to blow air (usually down) through the router body to cool the internal components.

Z Travel

When compared to the x- and y-axes, the amount of *z travel* on a CNC router is short (z travel is the total distance the z-axis can move, from its highest point to its lowest point). A typical setup allows for between 2 and 6 inches. As the height of the z-axis is adjusted, the cut depth changes. As the tool cuts deeper into the material, the *z* value becomes a negative number.

In older CNC mill setups, such as the classic Bridgeport mill shown in Figure 2-4, the z-axis has a much larger distance of travel than on a CNC router setup. This additional travel is to accommodate work fixturing, such as vises and fixture plates.

Figure 2-4. *Bridgeport mill*

An important z-axis technical specification is *z clearance*, or how much vertical space is available when the carriage is at its highest point. Three main factors determine the available *z* clearance: material thickness, the method used to attach it to the wasteboard, and the length of your cutting tool. If you need to cut thick material or if your project requires tall clamps or a long bit, your *z* clearance will be limited.

Table

The *table* is your work zone, the area where the raw material for milling will be mounted for cutting. The table comprises two components: the *platen* and the *wasteboard*. Let's take a look at each of these parts:

Platen
> The durable, flat surface under the cutting tool and other mechanical components. The platen is the "bottom" of the machine.

Wasteboard
> On top of the platen is the *wasteboard* (also called the *sacrificial layer* or a *spoilboard*), where you place the material to be cut.
>
> This sacrificial layer acts as a barrier between the bottom of your part and the platen, protecting both tooling and the platen. It is a *consumable* and will need to be periodically replaced or *resurfaced*.

When machining parts, you often cut all the way through the material, either to make profile cuts (as discussed in Chapter 9) or to remove the finished project from the scrap material. As you make these cuts, your bit will travel through your material and into the wasteboard, creating gouges, scratches, and scars.

In some cases, the platen is made from metal, and you don't want your precious router bits to hit the platen, damaging both your tooling and the machine. In most DIY machines, MDF is used as spoilboard because it is flat, strong, and heavy; has a consistent thickness; and doesn't tear out like plywood.

Mechanisms for Securing Materials

There are a variety of mechanisms you can utilize to secure materials to your table, including T-slot tracks, threaded inserts, double-sided carpet tape, wood screws, and clamps.

Imagine the forces of the machine while it's moving through your material. Because these forces are great, you need to implement a system to prevent them from moving your stock material. The following sections detail the various mechanisms for securing your materials.

Step Clamps

A common CNC accessory is a set of *step clamps*. These clever blocks allow a variety of thicknesses to be clamped with the same set of tools.

As shown in Figure 2-5, a step clamp is made up of several parts. There is a riser, the vertical member that looks like a miniature staircase. The arm is the horizontal member that has one end placed on top of the material and the other serrated end mated onto the riser.

Typically, the arm is slotted to allow a screw or bolt through access to the nut, typically inserted into the T-slots of the bed.

Step clamps come in a variety of heights and sizes; most kits include a small selection of several sizes.

When using step clamps, it is important to position the back of the clamp (the parts that mate with the riser) *above* the top of the material. By creating a slight downward angle, you are creating more downward force over the material (as illustrated in Figure 2-6). If the back end is lower than the material, most of the downward force from your bolt will be pushing down the riser and not the material!

Figure 2-5. Parts of a step clamp

Figure 2-6. Step clamp in action

T-slots

T-slots are an ingenious design that gives access to a slot from the end of the table. The slot itself is shaped like an upside-down *T* (hence, the name *T-slots*). All T-slots can be used two ways: you can either use a nut in the slot, and then thread a bolt into the nut, or a bolt can be placed in the slot (upside down), and then the nut run down over the threads to provide the clamping.

T-slots are typically used with step clamps, vises, or other clamping mechanisms. Figures 2-7 and 2-8 show the T-slot bed of a Tormach PCNC 1100 CNC mill.

Figure 2-7. *T-slots shown with T-nut and screw protruding upward from a Tormach PCNC 1100*

Figure 2-8. *T-slots (end view)*

Threaded Inserts

Threaded inserts are a fastening system that you have probably run into but may not have noticed. Threaded inserts are cylindrical and threaded on both the outside and the inside! Generally, you match up the type of material you are putting the threaded insert *into* with the type of threaded insert you purchase.

Screws

If your wasteboard is made from MDF, or a nonbrittle plastic such as HDPE or ultra-high-molecular-weight polyethylene (UHMW), the simplest and most straight-forward method for attaching your material is to just screw it down to the wasteboard. Yep, a screw in each of the corners of your material and then right into the wasteboard.

The downside to this method is that your wasteboard will eventually look like Swiss cheese! Each time you fasten a new piece of material to the wasteboard, you'll be contending with holes. Eventually, you'll need to replace your wasteboard. If it's made from commodity materials (like MDF), this shouldn't be an issue.

Use drywall screws and drill pilot holes in your stock material slightly *larger* than the size of your screw (but not larger than the screw head!). This will give you the best holding, without distorting the bottom of your material.

Tape

Sounds crazy to use something like tape to secure your material for milling, doesn't it? Maybe, if when you think of tape, you think of Scotch tape or painter's tape. But if you know about fixture tape, then you know it's not a crazy idea.

Fixture tape is a double-sided tape made specifically for this application. It has a lot of holding power, but somehow leaves very little residue. Other types of tape are known to work as well, tape such as 3M double-sided tape (many varieties) or even double-sided carpet tape.

Vacuum Table

One of the cooler options for securing your materials is a *vacuum table*. This is typically a hollow box with a perforated top, like the one shown in Figure 2-9.

A vacuum (could be a proper vacuum pump or something like a Shop-Vac) is hooked up to the table to create suction. When the stock material is set over top of the perforations, the suction pulls down on the material and secures it while being machined.

Figure 2-9. *Dan Spangler's ShopBot Desktop vacuum table*

Vises

Talk to a machinist about workholding, and they will undoubtedly mention a vise (as shown in Figure 2-10). In the industrial CNC world, using a vise is standard practice, mainly because these machines have high *z* clearance.

Figure 2-10. *Four-inch vises mounted on Tormach PCNC 1100 CNC mill*

For most hobby machines, two fundamental problems make vises impractical:

- There's not enough z clearance to accommodate vise height plus material height.
- Hobbyist machines are typically set up to cut sheet materials, which are too large to fit in a vise.

But if a vise works in your setup, awesome!

Routers Versus Mills

You may hear subtractive CNC machines referred to as *CNC routers* or *CNC milling machines*. These terms are often used interchangeably, but there are a few key differences. Let's take a closer look at each of these:

CNC mills
   A CNC milling machine like the Nomad 883 from Carbide 3D (shown in Figure 2-11) is designed for heavy-duty, high-precision work. Rigidity for accuracy and heavy metal machining is key, and speed is less of an issue.

Figure 2-11. *Nomad 883 CNC milling machine*

Mechanical Overview   35

CNC routers
  A CNC router like the Shapeoko (shown in Figure 2-12) is typically designed to cut *sheets* of material (like plywood or plastic). Sheet materials used on a router are generally softer and the cuts are less aggressive, so speed is preferred over rigidity.

Figure 2-12. *Shapeoko CNC router*

Machine Configurations

Most of today's DIY CNC routers and mills have standard but different x- and y-axis configurations, but there are some anomalies:

Moving table (y-axis), moving carriage (x-axis)
  Desktop CNC milling machines typically move the toolhead left and right along the x-axis, while the table moves forward and backward along the y-axis, as exemplified by the Nomad mill in Figure 2-11.

  This design provides the increased rigidity needed for industrial (and some DIY) mills to cut hard materials like steel and even titanium. Mills also provide more accurate cuts and are capable of much finer resolutions than routers for micro-machining softer substances. In addition, in many industrial mills, a vise is often used to secure a block of material.

Moving table (x- and y-axes)
In some cases, you will find a mill where the carriage moves up and down, while the table moves in both the x- and y-axes!

This is not a typical setup for most hobbyist CNC mills, but some (such as the LittleMachineShop 3501, shown in Figure 2-13) operate this way. One of the main advantages of a moving *x/y* table and a stationary z-axis is the increased rigidity of the machine because the cutting forces are always directly above the table.

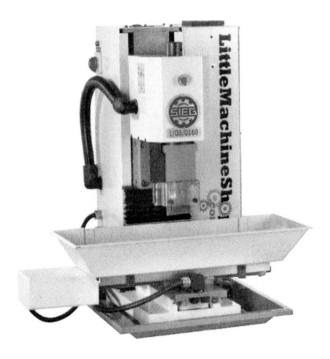

Figure 2-13. *LittleMachineShop 3501 CNC milling machine*

Routers: moving gantry (y-axis)
Most CNC routers (like Shapeoko) are designed to quickly cut sheet materials. Although they can usually cut soft metals (such as aluminum), they typically can't tackle hard metals (such as steel).

Mechanical Overview 37

For the remainder of this book, the focus is on *CNC routers* unless otherwise stated.

3/End Mills and Cutting

I've talked to hundreds of new CNC users over the years, and one of the most popular topics is CNC tooling, specifically *end mills* and how to use them. Fear not! Although these tool types may seem intimidating at first, there are only a few key features to understand before you dive in and begin experimenting.

End Mills

End mills are the sharp spinning tools used in CNC machining (Figure 3-1).

Figure 3-1. *An assortment of common end mills*

Drill Bits Versus End Mills

Although similiar in appearance to each other (Figure 3-2), an end mill and a drill bit are not the same thing, nor do they perform the same functions.

Figure 3-2. *Comparison of a drill bit (top) and an end mill (bottom) —note the different tips*

Drill bits

Drill bits are intended to be used to drill holes, plunging directly into the material at a 90-degree angle, as depicted in Figure 3-3. That's why the tip of a drill bit is pointed (as shown in Figure 3-2)—it's designed to bore straight down into the material.

The flutes (see "End Mill Anatomy" on page 45) on a drill bit are designed to guide the material being removed (known as *swarf*, or *chips*) up and out of the hole created by the bit.

End mills

As shown in Figure 3-4, end mills are designed to cut from their edges as they move laterally through the material.

A *center-cutting* end mill can both plunge into the material (like a drill bit) and move laterally across the material. End mills are designed to use their entire cutting edge length to cut, not just the tip.

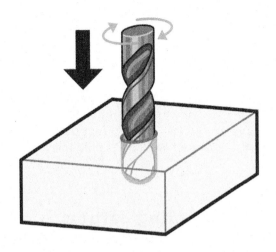

Figure 3-3. A drill bit cutting axially

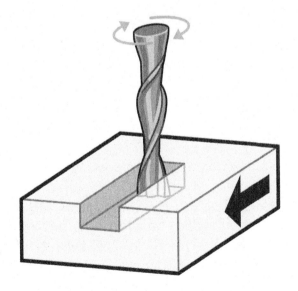

Figure 3-4. An end mill cutting laterally

Although most end mills are center cutting, some are not. If you're not using a center-cutting end mill, you'll need to either drill a pilot hole or begin all cuts outside the material envelope, and then move them into the material laterally. The non-center-cutting end mill is rare, so unless you're looking for one specifically, most of your cutters will be center cutting. If they're not, the product description from the store, catalog, or website should tell you.

> **Tooling Jargon**
>
> End mills are also commonly referred to as *bits*, *tooling*, and *cutters*. If you hear these terms in a CNC context, it's safe to assume the speaker is refering to end mills.

Common Tool Geometries

As illustrated in Figure 3-5, end mills come in a variety of profiles. The following are some of the most common types:

Straight flute
    Used to cut plastics such as HDPE or UHMW, or to tear through other soft materials.

Spiral upcut and downcut
    These are used for cutting hardwoods, softwoods, and plywoods (see "Chipload" on page 50).

   Spiral bits make clean edges when you are cutting. The spiral upcut bit pulls chips up and out of the slot being cut, whereas the spiral downcut bit pushes down as it's cutting.

   Because of this, with the upcut bit, you will get a very clean top edge, and with the downcut bit, you will get a fuzzy top edge but a clean bottom edge.

   Another type of bit is called a spiral compression bit. This is a combination of an upcut and a downcut bit. The top half of the cutting edge is designed like an upcut bit, and the bottom half of the cutting edge is designed like a downcut bit. When running this bit across the face of something like plywood, both your top and bottom edges will be clean.

Ball nose
These bits have a rounded tip, so unlike the straight end mills, at the end of a ball nose you will find a radius around the entire circumference of the cutter. These types of cutters are handy when rounded features or true 3D shapes are required. Because of their rounded tips, they do not make sharp inside edges where two faces meet.

Ball-nose end mills are not ideal for clearing large amounts of material because of their rounded ends. With the rounded ends, the amount of surface area the bit is engaged with, at its tip, is less than the diameter of the rest of the body. If you were to try to create a pocket with a ball end mill, the bottom of the pocket would have scallops. This can be avoided by cutting the overlap significantly, but ball end mills are designed to cut details and contours, not clear large amounts of material at a time.

V-Bit
Used to make signs or other detailed lettering applications. They are available in many sizes and angles, although the most common are 90, 60, and 30 degrees.

Engraving
Oddly shaped bits made specifically for shallow engraving and used most commonly for PCB creation. Although similar to the V-bit, engraving bits are intended for shallow cuts.

Table surfacing
Also known as a *fly cutter*, this specialty tool is made specifically for making very shallow depth cuts across a wide area. Fly cutters are typically used to surface the wasteboard on your machine.

Square (flat tip)
Great for removing large amounts of material and used for everything from roughing work to helical boring and finish passes. Their flat ends leave fewer surface toolmarks.

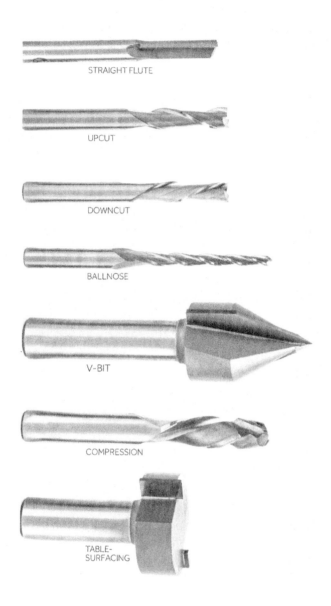

Figure 3-5. *An assortment of common cutting tools*

Tip Shapes

As shown in Figures 3-5 and 3-6 end mills have a variety of tips, each creating different cuts for a specific use.

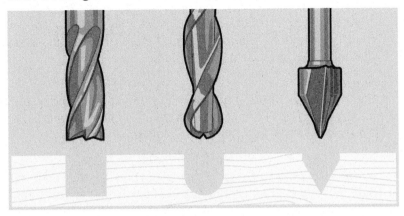

Figure 3-6. *Different tips create unique cuts*

If you are machining 3D contoured shapes that extend all the way to the bottom of a feature, using a ball end mill will result in a finer surface transition. It will also allow you to get the proper radius you desire closer to the bottom edges of your feature.

V-bits are used for engraving and even chamfering edges, but like ball end mills, they do a poor job of cutting profiles or milling out internal features such as pockets or profiles.

End Mill Anatomy

Let's briefly review the parts of an end mill (illustrated in Figure 3-7):

Overall length
   Measuring your end mill from one end to the other will yield the overall length of the bit.

Shank diameter
   This is the diameter of the part of the end mill that is not doing the cutting. In other words, this is the diameter that will go into your spindle. You'll need to know about your spindle's workholding capacity before selecting the Shank diam-

eter. Common shank sizes are 1/8 inch (3.175 mm) and 1/4 inch (6.35 mm).

Cutting edge diameter
This is the diameter of the cutting end of your end mill. If you were to plunge straight down into your material, and then measure the hole created, it would match the cutting edge diameter of your end mill.

Cutting length
End mills have flutes that cover only a portion of the overall length of the bit. The cutting length determines how deep of a pass can be taken with a single plunge.

Reach
Many end mills have a neck—a portion of the end mill between the shank and the cutting length that is not meant for cutting and is not meant for clamping. The reach length is the maximum distance your tool can go into the material.

Flutes
The grooves that spiral around your cutting tool. Each flute has a single cutting tooth that runs along the edge of the flute, from bottom to top.

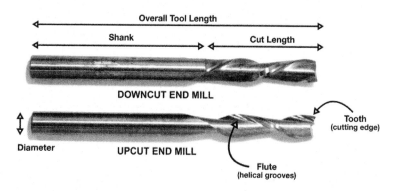

Figure 3-7. *Parts of an end mill*

End Mill Materials

End mills are typically made of either *high-speed steel* (HSS) or *carbide*. Carbide is harder, more rigid, and has a higher degree of wear resistance than high-speed steel. However, HSS end mills are generally less expensive, and because they are less brittle than their carbide alternatives, are less likely to break.

In most cases, for people just starting out, I recommend purchasing HSS bits because of their inexpensive nature. Once you're comfortable using your machine, and hopefully beyond accidentally smashing your bit into the table, move on to carbide. You'll find that carbide end mills last longer and give you a better finish.

Coatings

Most end mills (HSS or carbide) can be purchased with different coatings with crazy names like TiN, TiAlN, and TiCN. Coatings give end mills better wear resistance, cooling, and an overall longer lifespan. I recommend skipping the coating altogether until you really get into CNC.

Getting Started: What To Buy?

I recommend the following tools as a great starter set that can be used to make pretty much anything!

- Two-flute spiral upcut bit
- Two-flute ball nose end mill
- Straight flute bit
- V-bit (60 degree or 90 degree)
- Conical engraving bit

Cutting

As we discussed in "Drill Bits Versus End Mills" on page 40, most end mills are center cutting, meaning they can be plunged straight into the material, like a drill bit. However, this is hard on the machine, and it's hard on the tool. To minimize the stress on both your machine and your tool, and to reduce the likelihood of breaking a bit, most CAM software allows you to *ramp* into the material.

Ramping

As illustrated in Figure 3-8, ramping gradually increases the depth of cut as your tool advances through the toolpath. It's recommended that you use ramping for as many operations as possible. It takes a little longer, due to the lead in and gradual acceleration, but it's much easier on your machine and will yield more accurate parts and better tool life.

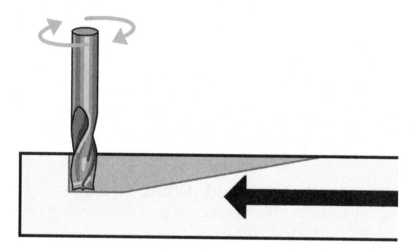

Figure 3-8. *End mill slowly ramping into the material*

Climb Versus Conventional Cuts

While running any given job, there are only a few times when the bit is completely engaged with the material, usually when the end mill plunges into the material. The rest of the time, the bit is working only to widen an existing cut. When the tool is not drilling or ramping, there are two path direction options: conventional and climb (Figure 3-9). Let's take a closer look at each of these:

Conventional cut
    This path is easier on the machine and can be used with less rigid machines. Because the chipload starts at nearly zero as the bit gradually works its way into the material, this strategy provides a better surface finish on most materials and a higher degree of dimensional accuracy.

Climb cut
Although more strenuous on the machine and the tool, climb milling can be faster, and in some materials creates a better surface finish due to better heat dissipation.

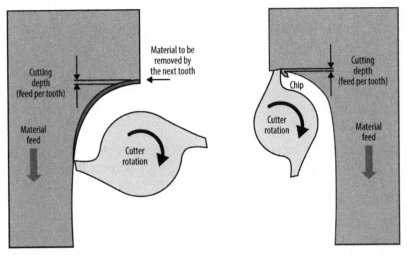

Figure 3-9. *Climb and conventional cutting strategy*

Speeds and Feeds

The phrase "speeds and feeds" refers to how quickly the spindle is turning, combined with how fast the tool is moving through the material. These measurements are communicated in units over time, commonly in *inches per minute* or *millimeters per minute*.

Don't Be Intimidated!

Understanding speeds and feeds is the most daunting CNC machining topic for beginners, but don't let it prevent you from getting started! It's not necessary to become a feeds and speeds expert. In fact, you can make a great many projects knowing only the bare minimum.

The following are some key terms when talking about speeds and feeds.

Feed rate
    Refers to how quickly the machine is moving through your material (laterally, not plunging, as shown in "Drill Bits Versus End Mills" on page 40).

Spindle speed
    Refers to the *revolutions per minute* (RPM) of your spindle.

Depth pass
    A key factor in determining your speeds and feeds is *depth pass*, or how deep the bit plunges into the material in a single pass. Another term for depth pass is *depth of cut*, or DOC for short.

When determining speeds and feeds, a good rule of thumb is to base the depth pass on the diameter of your cutter. For soft materials (like plywood), your depth pass is typically equal to the diameter of your cutter.

For denser materials (like hardwood), the depth pass should be about half the diameter of your bit.

When cutting really hard materials (like aluminum), depth pass should be a fraction of the diameter of your tool.

Chipload

As the flute's cutting tooth bites into the substance being machined, it removes a small chunk, or *chip*, with each pass. *Chipload* is the thickness of that chunk.

When an upcut end mill is used, the chip follows flute direction, until it is ejected at the top of the material, as shown in Figure 3-10.

 End mill manufacturers (like Onsrud (*http://www.onsrud.com/xdoc/FeedSpeeds*)) do a lot of the speeds-and-feeds dirty work for us by specifying the optimal chipload for any given tool they produce.

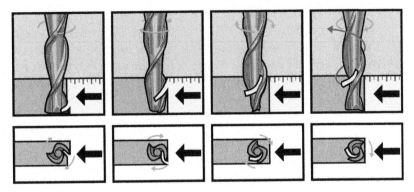

Figure 3-10. *Chip creation as tool moves through material*

CNC Math

The following formulas give you a starting point for determining the best speeds-and-feeds settings for any given material.

Calculating Chip Load
    Chip Load = Feed Rate / (RPM × Number of Flutes)

Calculating Feed Rate
    Feed Rate = RPM × Number of Flutes × Chipload

Calculating Spindle Speed
    Speed = Feed Rate / (Number of Flutes × Chipload)

Types of Tool Holding

The following is a list of a few common types of tool holding that you'll run into on most hobby grade CNC machines.

Jaw chuck
    Used on most cordless drills and used on drill presses. A jaw chuck needs a chuck key to tighten the jaws around the tool.

    This is a terrible way to hold an end mill. Jaw chucks have a large amount of runout and won't yield good results. If your machine uses a rotary tool, don't use a jaw chuck!

Fixed-diameter collet

With a fixed-diameter spindle, a metal sleeve is used to hold the tool. This sleeve is usually accompanied by a set screw to keep the bit in place. Fixed-diameter spindles are found on small machines with a 1/8-inch (3.175 mm) or less tool capacity.

 Your collet must match your spindle—a Bosch Colt router has a collet made for a Bosch Colt, a Dewalt DWP611 has a collet made for a Dewalt DWP611, a Dremel has a collet made for a Dremel. You get the idea.

ER collets

The best and most popular option for many CNCs, *ER collets* are designed to collapse around the cutting tool as they are tightened. They have a 1 mm range of clamping and both metric and Imperial bits can be used.

ER collets come in standard, number-coded sizes. The standard size for desktop CNC mills are ER-8, ER-11, ER-16, and sometimes ER-20.

If you're using a "proper" CNC spindle, odds are the spindle will use an ER series collet (probably ER-11 or ER-16). If you want to use different shank diameters with your ER-11 spindle, you'll need corresponding diameter ER-11 collets to match. They are often sold in sets, like the one shown in Figure 3-11.

Figure 3-11. *ER-16 collet set*

 The collets in Figure 3-11 are all the same external size, the only difference being *the bore* (or center hole) that matches your tool diameter.

Collets seem to intimidate people—they're confusing and most people brush over them. Have no fear! Let's briefly review what you need to know about collets.

Think of the collet as a gripper. The shape and size of the outside of the collet is specific to the spindle (as we noted earlier, Bosch, Dewalt, and the ER series are not compatible with each other, but they are all collets).

The collet has a hole (or a bore) in the middle that is a specific size. Let's say our collet has a 1/8-inch bore. The only thing we can put into that collet is a cutter with a 1/8-inch shank.

To make this collet work, it needs a nut, or more specifically a collet nut. The collet nut has a ring on the inside that allows the collet to snap into the inside of the nut.

End Mills and Cutting

Once the collet is inside the nut, you can slide your cutter into the bore. Then you can insert the nut and collet into the spindle.

The inside of the collet nut is also threaded. When you begin threading this onto the end of your spindle, the collet will be pushed inside the spindle. As you continue to tighten the collet nut, the collet will be pushed farther up the spindle until the side of the collet engages with the inside of the spindle neck. Once this happens, as you continue to tighten the nut, the collet, having nowhere to go because it's trapped in the neck of the spindle, will be pushed in toward its center.

Because of the way it's designed, the collet begins to act like a spring and the walls start to collapse toward its center. As the wall collapses, the bore of the collet will begin pushing (from all sides) onto the cutter. When the collet nut is tightened all the way, the collet bore will have been reduced to slightly smaller than the diameter of the cutter shank. Because of this, the cutter will be trapped inside the collet.

4/CAD: Draw or Model Something

In the wonderful world of CNC, every project begins with a digital design. If you're already familiar with CAD, or *computer-aided design* software, then you're already well on your way to getting started with CNC. If you're new to drawing on a computer, don't worry—it's easier than you think—and there are dozens of free options available.

While it's entirely possible to machine 3D models, most CNC machining projects are designed in two dimensions using CAD or *vector* graphics software. Then the magic of CAM software allows you to assign separate operations to individual shapes within your drawing. (We'll discuss machining 3D models and using CAM software in Chapter 5.)

CNCs use *computer-aided manfacturing* (CAM) software to process digital files through the typical workflow outlined in Figure 1-2. CAM programs will accept only specific file types. The range of formats can vary, but the overall principles are the same.

While both *raster* and *vector* images are two-dimensional, rasters are made of *pixels*, while vector drawings are made of math. The following section discusses raster images, and we'll cover vector drawings in "2D Vector Graphics" on page 57.

2D Raster Images

A raster image, also called a *bitmap* when it is only two colors (usually black or white) or monochrome, is made up of colored squares, or pixels. Pixels (short for picture element) are the smallest unit that can be displayed on a screen. The level of detail the images are filled with is referred to as *pixels per inch* (PPI).

Good-quality photographs typically require a 300 PPI *resolution* (image width, height, and pixel count). This means that for every 1-inch × 1-inch area of a full-color, high-resolution pixel image at 300 PPI, there are 90,000 pieces of color information!

Because rasters are composed of colored squares, they're pretty terrible at scaling up. As the image in Figure 4-1 is enlarged, the pixels themselves become larger, reducing your PPI. That's why raster images look blurry, or *pixelated*, when stretched beyond their native resolution, as shown in Figure 4-2.

```
RASTER
IMAGE
```

Figure 4-1. *Raster image at native resolution*

Figure 4-2. *Scaled-up raster image*

 Common pixel file extensions include JPG, JPEG, BMP, PNG, and TIFF.

2D Vector Graphics

Vector graphics use *paths* connected by points, or *nodes*, to define shapes that describe the shape and proportion of a design, as illustrated in Figure 4-3. Unlike bitmaps, vectors are easily manipulated and scaled, as shown in Figure 4-4.

Remember the 1-inch × 1-inch raster image that had 90,000 pixels, each filled with a solid color? The same square drawn using vectors would be made of four points (one for each corner), with mathematically generated paths (also called *strokes*) connecting each point. *Fill* is the commonly used term for the color or texture that applies color inside the square's paths.

CAD: Draw or Model Something   57

Figure 4-3. *Vector, resolution increased, clarity preserved*

Figure 4-4. *Vector image scaled up 400x to show lack of distortion*

 Common vector file extensions include SVG, AI, EPS, PDF, CDR, DXF, and DWG.

Vector Editing Software

There are many vector editing programs. Some are designed for creating artwork, while others are digital replacements for manual drafting used in architecture or engineering. Common packages include the following:

Art-based programs
Inkscape (*http://www.Inkscape.org*), Adobe Illustrator (*http://www.adobe.com*), iDraw (*http://www.indeeo.com*), and CorelDraw (*http://www.coreldraw.com*)

Drafting-based programs
AutoCAD (*http://www.autodesk.com*), LibreCAD (*http://www.librecad.org*), QCAD (*http://www.qcad.org*), and DraftSight (*http://www.3ds.com/products-services/draftsight-cad-software/*)

Inkscape

Released under the GPL, Inkscape is a free and open source *scalable vector graphics* (SVG) editor with a mature feature set that rivals commercial programs such as Adobe Illustrator and CorelDraw (Figure 4-5).

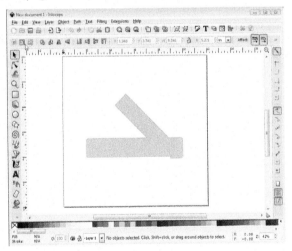

Figure 4-5. *Inkscape software interface*

To follow along with the design examples in this book, download and install Inkscape (*http://inkscape.com*), or open your vector editor of choice.

2D Drawings Versus 3D Models

To demonstrate the differences between 2D and 3D file types, we'll walk through how to create the same push stick in each format.

 Unlike 3D printing, where a 3D model is needed, basic 2D designs are more commonly used in CNC machining, particularly by beginners. In both this chapter and Chapter 5, I'll keep the focus on 2D design and machining, with brief asides on 3D.

To demonstrate how to think through a two-dimensional vector drawing, I'll walk you through the overall design process.

First, you'll need a project. What do you want to create? You may find it useful to sketch your idea on paper, as shown in Figure 4-6. As you draw, include important design details, such as dimensions and angles.

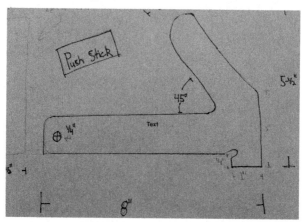

Figure 4-6. *Push stick sketch*

After finishing your paper sketch, open Inkscape (or your vector editor of choice) to begin re-creating your sketched push stick idea digitally. Figure 4-7 shows my push stick vectors in Inkscape.

Although teaching you how to use Inkscape is outside the scope of this book, Inkscape's dedicated user community has created tons of written tutorials (*https://inkscape.org/en/learn/*) and video walkthroughs (*https://inkscape.org/en/learn/videos/*) to get you started.

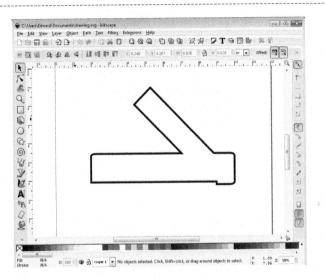

Figure 4-7. *Push stick vectors in Inkscape*

After the design has been drawn, save it in a file format that can be opened with your CAM program, which is where you will create the toolpaths (we'll discuss CAM in more depth in Chapter 5).

 Inkscape can save files as SVG, PDF, or even DXF, which are all acceptable input file formats for 2D machining.

3D Models

In order to re-create that project in 3D, you'll need to use 3D modeling software.

The result would look something like Figure 4-8.

Figure 4-8. *3D push stick model*

3D Modeling Software

3D modeling software packages vary widely and range from free to extremely expensive. Here are a few popular options to get you started:

Free software
  OpenSCAD (*http://www.openscad.org*), Blender (*http://www.blender.org*), and FreeCAD (*http://www.freecadweb.org*)

Commercial software (free versions)
  Onshape (*https://www.onshape.com*), SketchUp (*http://www.sketchup.com*)

Commercial software (paid versions)
  Autodesk Fusion 360 (*http://autode.sk/2ahXyTn*), Autodesk Inventor (*http://autode.sk/2aRjzcW*), SolidWorks (*http://www.solidworks.com/*), and Rhino (*http://www.rhino3d.com*)

More Software to Try

There are several software packages available that generate G-code for specific tasks such as V-carving (engraving with a V-bit), converting rasters, single-line drawing, and halftoning. I've listed a few of my favorites in the sections that follow.

V-Carving Text

There are programs made specifically for V-carve engraving text. Scorch Works offers an open source program for this called F-Engrave (shown in Figure 4-9), which can be downloaded at its website (*http://bit.ly/2ayHY2h*). F-Engrave will also accept a DXF and will auto-trace an image!

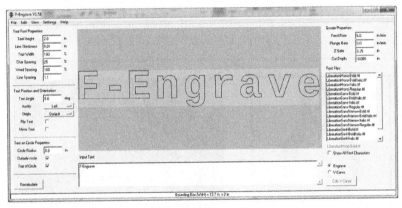

Figure 4-9. *F-Engrave interface*

Image to G-code

Converting images to G-code is an easy way to turn nonvector art or even designs into G-code without the need to re-create the original. The other benefit of image to G-code is it lends itself nicely to nonengineered designs, such as photographs and artwork.

Images are converted to G-code by creating a depth map of the image. A depth map is a grayscale image in which the darkness or lightness of each pixel corresponds to the depth of the object at each point.

We know that raster images are simply pixels turned on and off or with different colors. Depth maps read the value of each pixel and set a corresponding depth dimension: typically, darker objects are *deeper*, and lighter objects are *not as deep* (or the reverse).

Quite a few software packages are available for download that perform these tasks. The following are two of my favorites:

- image2Gcode (*http://www.thuijzer.nl/image2gcode*), for a web-based option
- dmap2gcode (*http://www.scorchworks.com/Dmap2gcode/dmap2gcode.html*), which is available for Windows and Linux

Figure 4-10 shows the before and after of using an image of Linux mascot Tux to generate a toolpath using image2gcode.

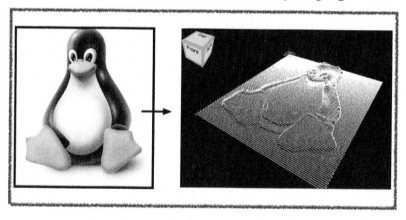

Figure 4-10. *Using image2gcode to create a toolpath for an image of Tux*

Single-Line Drawing

A fun way to pass the time with a CNC machine is to create line drawings, sometimes referred to as single-line drawings. You can read the blog post "StippleGen: Weighted Voronoi Stippling and TSP Paths in Processing" (*http://bit.ly/2a6F3Pm*) to learn more about them.

Single-line drawings are exactly what their name indicates: a drawing made from a single line! As you can see in Figure 4-11, single-line drawings are similar to those you might see on an Etch A Sketch toy.

One of the cooler parts about watching your CNC machine do a single-line drawing is how fast it can run. Because the z-axis never raises after the initial engagement, the machine runs in only the x- and y-axes. Because of this, and because there are little to no forces put on the machine, you can generally run much faster than if you were cutting through a material.

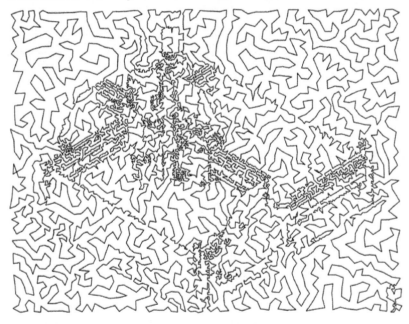

Figure 4-11. *Single-line drawing of a Shapeoko v1*

For a full tutorial on single-line drawings, see this walkthrough on the Shapeoko wiki (*http://www.shapeoko.com/wiki/index.php/Tsp_sld*).

Halftone Images

One of my favorite things to do with a CNC machine is to convert an image into G-code via halftoning the image (*http://jasondorie.com/page_cnc.html*) (see Figure 4-12).

Figure 4-12. *Grayscale image converted to halftone and machined*

After ideas have been drawn digitally, the next step is to make the toolpaths for all of the features in your design.

5/CAM: Make Toolpaths

Computer-aided manufacturing, or *CAM*, programs are used to define *toolpaths*—the roadmap a CNC follows to physically cut a digital design file.

Toolpaths are (typically) created by importing a 2D CAD or vector graphic or a 3D model into a CAM program, and then applying *operations* and *parameters* to that file. Parameters are user-entered values, such as the diameter of your cutter, speeds and feeds, and cut depth. These values are applied to operations, or cutting types, creating the instructions (made up of coordinates), that tell the machine how to move and cut.

Like 2D and 3D design software, 2D and 3D CAM programs are completely separate software packages. They import different file types and have different features and operations.

DIY CAM Software Packages

For a free 2D CAM software package, try MakerCAM (*http://www.makercam.com*). The following commercial options are also available:

- CamBam (*http://www.cambam.co.uk*)
- Carbide Create (*http://carbide3d.com/carbidecreate/*) (included with Carbide 3D CNC machines)
- Carbide Motion (*http://www.carbide3d.com*)
- Vectric (*http://www.vectric.com*) (Aspire, Cut2D, VCarve, PhotoVCarve)

> For a free 3D CAM software package, try PyCAM (*http://pycam.sourceforge.net*). The following commercial options are also available:
>
> - Autodesk Fusion 360 (*http://autode.sk/2ahXyTn*)
> - MeshCAM (*http://www.grzsoftware.com*)
> - VCarve Desktop, VCarve Pro, Aspire, Cut3D (*http://www.vectric.com*)
> - DolphinCAM (*http://www.dolphincadcamusa.com*)

As you begin to look at CNC projects with a critical eye, you'll recognize that *true* 3D machining is relatively rare. Most CNC'ed parts are designed in 2D, then machined in either 2D or 2.5D.

2D/2.5D Toolpaths

Both 2D and 2.5D toolpaths can be created from any (or even the same) 2D drawing. The difference is that 2D toolpaths cut only at a single depth, while 2.5D cuts at multiple depths.

An easy way to remember this is to think of a pen plotter. The pen plotter is *cutting* a 2D design. The pen goes down onto the paper, and the machine moves in only the *x* and *y* directions. This is technically 2D.

2.5D means that the part has more than one thickness. For instance, think of a kitchen cutting board for chopping and preparing food: the board may have a slot cut in the top that goes the entire thickness of the board—this serves as a handle. The board may also have a shallow channel cut just outside the perimeter of the cutting area. This channel prevents liquids from spilling over the edges.

The channel makes the cutting board 2.5D. It is a feature that is cut into the board, but does not go all the way through.

In some edge cases, the line between 2.5 and 3D gets blurred a little. Both V-carving and ramping into cuts are technically 3D machining, as they are moving all three axes at the same time. However, those are more a built-in feature to the CAM software than they are a description of the part geometry.

3D Toolpaths

3D CAM programs also cut at multiple depths, but they create the contoured surfaces they produce by generating toolpaths from 3D models. Cutting 3D files results in moving the x-, y-, and z-axes simultaneously.

At some point in your CNC journey, you'll find a need for all three of these types of designs. I suggest beginning with 2D, then experimenting with 2.5D, until your project requires an actual 3D model.

2D/2.5D CAM Operations

A typical 2D CAM package has at least five basic 2D/2.5D cutting operations: *outside profile*, *inside profile*, *pocket*, and *engrave*, as shown in Figure 5-1.

As mentioned in "DIY CAM Software Packages" on page 67, the difference between 2D and 2.5D machining is depth of cut. 2D CAD software does not automatically specify cut depth and operation type. The same 2D file can be used to create several different toolpaths. You could use a *profile* cut to go all the way through, the middle could be machined at 50% material thickness or cleared out as a *pocket*, and the surface could be engraved.

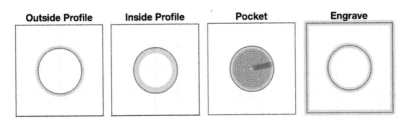

Figure 5-1. *Common 2D and 2.5D CAM operations*

In Figure 5-1, the 2D vector design is the dark red line. The red shading simulates how the machine will move as it cuts the file and the width of the material removed.

The amount of material removed (and the size of the shaded area) is determined by the diameter of the end mill used.

> ## Offsetting: Compensating for Kerf
>
> *Kerf* is the gap made by cutting with a saw blade—or an end mill. As the cutter follows the toolpath through the material, it leaves a path (as wide as the bit) where there is no material left.
>
> To compensate for kerf and cut dimensionally accurate 2D parts, the end mill needs to be *offset* from the vector by half the diameter of the tool.

Let's briefly review some of these key terms:

Outside profiles
   Move around a path, keeping the tool offset to the *outside* of the path. An outside profile cut is typically used to cut a part away from the material.

Inside profiles
   Cut features *inside* the bounds of a part.

 If your project is a picture frame, you would use an outside profile to cut the outside of the frame, and an inside profile to cut the window for the picture.

Pocket
   When a feature located inside the bounds of that part is "cleared out" to a specific depth. Pocket cuts do not go all the way through the material.

Engrave
   The center of the bit is placed directly on the line of the feature, tracing your shape's path.

Peck drilling
> Mimics how a person would operate a manual drill press. The bit plunges straight down a short distance into the material, then retracts so the chips can clear. This process is repeated until the target depth is reached.

2D/2.5D Toolpath Parameters

Every toolpath must be assigned settings, or *parameters*. These parameters tell the machine how deep to make each cut, how much overlap should occur between passes, and safe operating depths/heights for your particular machine. Each CAM package is a little different, but the MakerCAM screenshot shown in Figure 5-2 is a good example of a typical dialog box for settings.

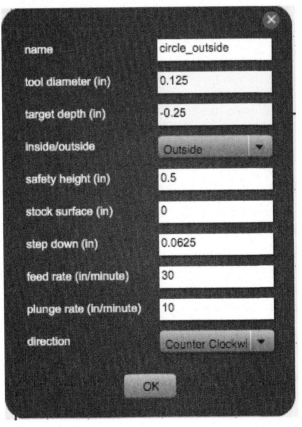

Figure 5-2. *2D parameters, shown in MakerCAM*

Mowing the Lawn

Have you ever mowed a lawn with a push mower? If so, then you've already created toolpath-like patterns that incorporated *max depth*, *depth per pass*, and *stepover*. Think of the entire process as a CNC machine setup. Lawn mower blades are the spindle, the mower width is the diameter of the bit, and the lawn is the stock material.

When I was a boy, my dad taught me how to achieve a constant, even mow. I'd cut a path to the end of a row, then turn the mower around, overlapping (or "stepping over") the freshly cut, neighboring parallel path by about 6 inches.

If you've just arrived home from vacation (and the lawn hasn't been mowed in weeks), you shouldn't attempt to cut the grass to the desired height in one go—you'll clog up and stall out your mower! The same thing happens to your spindle when you use a *cut depth* that's too deep.

To keep the mower from clogging, you'd cut the overgrown yard in two (or more) stages, reducing the amount of grass removed in each *step-down* pass. It's the same with CNC machining—keep your depth pass estimates conservative and be kind to your machine.

Let's briefly review some of these key terms:

Tool diameter
    The diameter of the tool to be used.

Target depth
    This number defines how deep you intend to cut the feature. If you intend to cut all the way through the material, the target depth should very slightly exceed the material thickness.

Operation
    How to cut the vector: inside, outside, or on the line.

Step down
    The depth of material removed during a pass. Also referred to as *cut depth* or *depth per pass*.

Stepover
> Toolpath overlap is called *stepover*. If you increase your stepover, machined cuts will look cleaner and smoother. Increasing your stepover will make your job take longer to cut, but results in a better finish.

Feed rate
> How fast your tool moves across the material when cutting (see "Speeds and Feeds" on page 49).

Plunge rate
> How fast your cutter moves *down* into the material.

Direction
> Specifies climb or conventional cutting (see "Climb Versus Conventional Cuts" on page 48).

Safety height
> Sometimes called *safe z* or *clearance height*. Between cutting moves, the z-axis will raise above the workpiece in order to travel to the next operation's starting position.
>
> The safety height defines how high above the material to raise before moving to the next point. This should be higher than any clamps you are using.

 Order Matters!
Toolpath order is important, and you'll need to think it through carefully while creating the files. All pocket, drilling, or engraving operations must happen first, prior to cutting all the way through the material. Once you cut your part free of the stock material holding it in place, it's no longer possible to perform additional operations.

Overcuts

It you're attempting to mate two pieces with sharp outside corners but rounded inside corners, as shown in Figure 5-3, then you'll run into trouble.

The "Inside Corner" Problem

The inside corners of profile toolpaths have rounded inside corners. This occurs because the rounded outside of the bit is not able to make it all the way into the corner of the feature.

Think of tracing the inside wall of a cardboard box with a can of soup. The can makes direct contact with the box walls, but its round shape is unable to completely fit into the tight corners—there's always a gap.

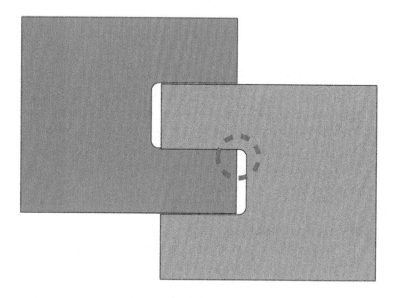

Figure 5-3. *Rounded internal corner collision*

Although the length and width dimensions on the inside profile will accommodate those of the mating piece, those pesky rounded corners keep the parts from fitting together perfectly!

You could use a sharp chisel to make the round inside corners square, but it's more appropriate to use CNC joinery techniques so the machine does the precision, press-fit work for you.

The most common method of designing around the "rounded corners" limitation is to create *overcuts* in your CAD file. The most common overcut types are known as *dog bones* and *T-bones*.

Dog Bones

In a dog-bone overcut (Figure 5-4), the inside profile corner is enlarged into a semicircle, allowing the sharp edges of the mating piece to slide through without interference.

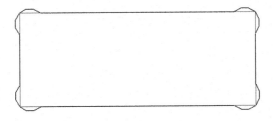

Figure 5-4. *Overcutting with dog bones*

When a dog bone is applied to the inside corners, as shown in Figure 5-5, the joints fit together nicely!

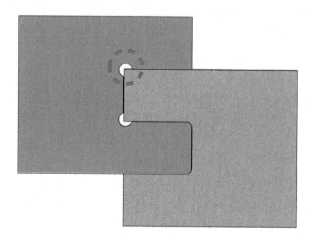

Figure 5-5. *Dog-bone overcuts create a perfect fit*

T-Bones

The T-bone overcut design is very similar to the to the dog-bone design, but the placement of the corner enlargement differs slightly, as illustrated in Figure 5-6. However, the effect is the same—it creates extra room at the corner to allow the square mating parts to slide through.

Figure 5-6. *Overcutting with T-bones*

Vectric (and some other CAM software) has a built-in *fillet* tool that can automatically insert dog- or T-bone overcuts into internal profiles.

Minimum Feature Size

When creating toolpaths in a CAM program, it's vital to simulate your job to ensure that your smallest design feature will be machined. The minimum feature size dictates what size cutter you'll need to achieve the effect you want.

Figure 5-7 shows pocket, inside profile, and outside profile toolpaths (in blue) that have been generated for a digital design using a 1/4-inch square end mill.

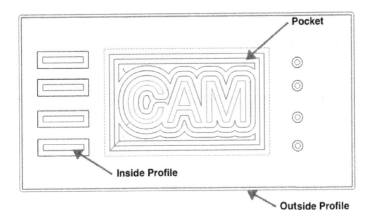

Figure 5-7. *Toolpaths applied*

Next, take a look at Figure 5-8. The gray shading shows where the tool will travel to remove material. The white indicates the material that will be left after the file is cut.

The white area in the inside the letter *A*, between the letters *C* and *A*, and the legs of the *M* will not be machined.

Selecting the Right Tool

As illustrated in "Tip Shapes" on page 45, end mill tips (like V-bits and ball-nose cutters) make different-shaped cuts. Many CAM programs take the tool geometry as a parameter and show the result in the simulation preview. When creating toolpaths for your parts, be aware of the profiles your tools will be creating.

The letters in Figure 5-8 won't be properly cut because the 1/4-inch end mill selected was too large and could not fit into the minimum feature size between the letters.

Figure 5-8. *Material in letters is not completely removed, because tooling is too large*

To get in and machine those features, you need to use a smaller bit. If the toolpath is re-created using a 1/8-inch diameter tool (Figure 5-9), the letters become disconnected and more easily visible.

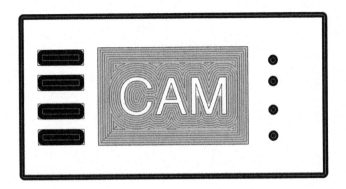

Figure 5-9. *A smaller end mill can fit into the features*

 There's still a slight radius in the corners, but that could be eliminated by using a V-bit or an end mill so small that the rounded corners become invisible.

Basic 3D CAM Operations

As mentioned previously, with 3D toolpaths all of the machine's axes (x/y/z) can all be cutting at the same time in synchronized motion.

3D CAM packages have similar settings to 2.5D in terms of job setup and operations. The most significant difference is the strategy used to machine the part. Most 2.5D and 2D programs require the user to select each feature and then apply the appropriate toolpath, but with 3D, the user selects a strategy.

Common strategies include parallel and contour finishing.

Parallel Finishing

Parallel finishing is a very popular strategy to use with parts that do not require a great deal of change in depth across their plane (see Figure 5-10). The toolpath moves the machine parallel to either the x or y plane. Depending on your CAM program, different combinations may be used, including zigzag, "x only," or "x and y."

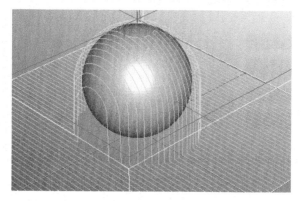

Figure 5-10. *Parallel finishing toolpath strategy generated by MeshCAM*

Contour Finishing

Contoured finishing strategy creates a toolpath with a constant z step down (see Figure 5-11). This means the part will begin being cut at its highest point, and the cutter will follow the contour down to the base of the machine. The advancement in the z direction will be dictated by the step-down value.

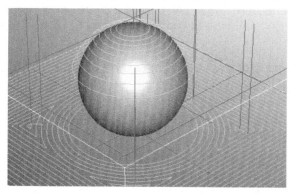

Figure 5-11. *Contoured finishing toolpath strategy generated by MeshCAM*

6/CAD/CAM Project: No Machine Necessary!

This chapter unites the CAD/CAM process by outlining the steps needed to design a wooden toy racer, then create and simulate the toolpaths—even if you don't own or have access to a CNC.

Inkscape

Inkscape (*http://inkscape.com*) is a free and open scalable vector graphics editor (see "Inkscape" on page 59). If you didn't download as you followed along in Chapter 4, go ahead and do that now.

MakerCAM

MakerCAM (*http://www.makercam.com*) is a free, fully functional, browser-based CAM software with basic built-in CAD functionality (see Figure 6-1). You can draw circles, rectangles, rounded rectangles, and polygons, and do basic vector editing.

MakerCAM can import an SVG file, create toolpaths, and save those toolpaths as a set of machining instructions or a G-code file (explained in detail in Chapter 8) to your local computer.

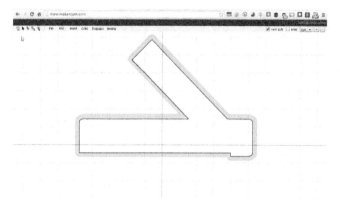

Figure 6-1. *MakerCAM screenshot of the push stick*

 MakerCAM isn't perfect—it has design limitations that may cause some people to outgrow its functionality at some point. However, it's free and works rather well, so it's the perfect CAM program for getting started if you don't already have a solution.

Webgcode

My favorite free and open G-code visualizer is Nicolas Raynaud's Webgcode (*http://nraynaud.github.io/webgcode*) (shown in Figure 6-2).

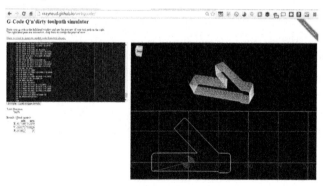

Figure 6-2. *Webgcode G-code visualizer*

CAMotics

If you are looking for a standalone application, take a look at Joseph Coffland's CAMotics (*http://camotics.org/*) program (Figure 6-3). It's a really powerful cross-platform simulator that lets you create your own tool library and actually *watch* the G-code as it's translated (virtually) into motion.

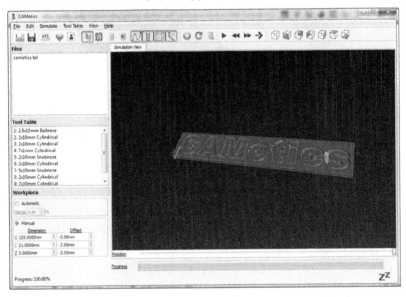

Figure 6-3. *CAMotics G-code visualizer by Joseph Coffland*

The best way to become a better CNC operator is to practice. By using the software workflow outlined in this project, you can practice designing and creating toolpaths for CNC without ever actually touching a CNC machine!

Visualizers give you the ability to preview your toolpaths before you perform the work. Regardless of your skill level or access to a CNC, previewing your toolpaths before running the G-code on your machine is *always* an excellent idea!

Wooden Racer Project

The wooden racer (Figure 6-4) is a small desk toy. To keep things simple, the design uses a stock piece of wood and commonly available hardware.

This 2D example project is basic, but the techniques used here could be applied to many projects. The tools I've selected are both free and open, and I still use them often for certain projects.

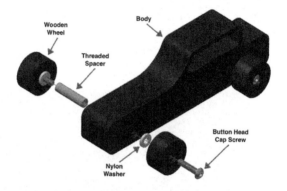

Figure 6-4. *Wooden racer assembly diagram*

 Even if you're not physically machining and building the project, it is still helpful to note the part dimensions when thinking through and creating your CAD file.

Project Materials and Dimensions

If you have a CNC (or CNC access) and are able to build the wooden racer, you'll need some wood and the following components from McMaster-Carr (*http://www.mcmaster.com/*):

A piece of wood (or other material)
   It should measure at least 8 inches wide, 6 inches long, and 1 inch thick.

#8 threaded spacer
- McMaster-Carr (PN 96110A034)
- Outside diameter: 0.25 inches
- 1 inch in length
- 8-32 screw size

#8 button-head socket cap screw
- McMaster-Carr PN: 98164A139
- 8-32 thread
- 3/4-inch length

#8 nylon washer
- McMaster-Carr PN: 90295A400
- Inside diameter: 0.173 inches
- Outside diameter: 0.375 inches
- Number 8 screw size

Step 1: Create the Digital Design

Use Inkscape to turn your sketch into a digital drawing. Feel free to experiment—you don't need a machine to create toolpaths. Practice simulation as much as you like; it's good to become familiar with the toolpath and G-code generation process before machining.

Body

I found it easiest to design the racer body with the Inkscape path tool and just approximate the shape and contour of the design. The only real requirement is that the body has two holes near the base for the axles, as shown in Figure 6-5.

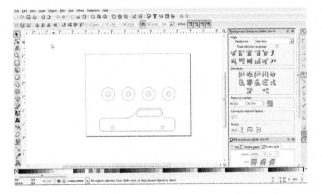

Figure 6-5. *Wooden racer drawn in Inkscape*

If you'd like to skip the file creation and get right to the CAM and simulation, download (http://DOWNLOAD-LINK-TBD.html) the Inkscape wooden racer file.

Wheels

Inkscape has several built-in geometric shapes. Use the circle shape to create four wheels, each made up of three concentric circles, shown in Figure 6-6.

Figure 6-6. *Wheel concentric circles*

As discussed in "DIY CAM Software Packages" on page 67, 2.5D designs (features machined to different depths, like pockets), are created in 2D software, then assigned different operations and parameters in CAM software.

You'll select the feature for each toolpath operation, and specify depths for each feature. The end result will look something like Figure 6-7.

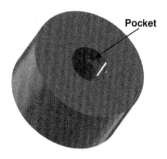

Figure 6-7. *Wheel pocket and profiles for screw and axle*

When you're reasonably happy with the design, save the file in the SVG format as *racer.svg*.

Step 2: Configure MakerCAM

Before importing the *racer.svg* file into MakerCAM, you'll need to change a few MakerCAM settings to ensure the design is imported at the correct scale:

1. Open the Preferences menu (Edit → Preferences).
2. Change "SVG Default Import Resolution (px/in)" to 90, as shown in Figure 6-8.

Figure 6-8. *MakerCAM SVG import settings*

Step 3: Import and Center Racer SVG File

Close the Preferences window and import the *racer.svg* file.

Your design will probably import into the upper-right part of the work area, because in Inkscape, the bottom-left corner of the canvas is considered X0, Y0, and everything you drew was "up" and to the "right" of that point.

To position your part in the middle of the workpiece, select the cursor tool in the top left of the interface and highlight all the vectors. Then drag them around until they are centered on the origin, as shown in Figure 6-9.

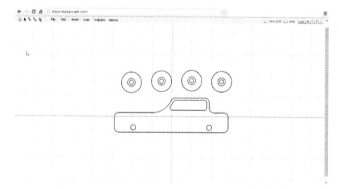

Figure 6-9. *Wooden racer imported into MakerCAM and centered*

Now comes the exciting part: creating the toolpaths for your design's individual features! Figure 6-10 outlines each of the toolpaths you will be creating.

Step 4: Create Wheel Toolpaths

After several simulations, creating toolpaths will begin to come naturally and you'll begin thinking about how the actual cuts will look as you are designing your parts.

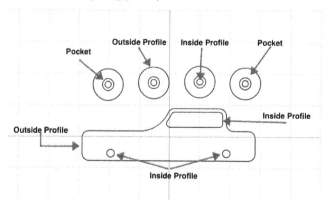

Figure 6-10. *Overview of toolpath operations used in the wooden car design*

Reduce Wheel Thickness by Half

The wood is 1-inch thick, but the design calls for 1/2-inch-thick wheels. To reduce the wheel thickness, follow these steps:

1. Select the wheel's exterior circle.
2. Next, you need to apply a pocket operation. Navigate to CAM → Pocket operation.
3. As shown in Figure 6-11, set the target depth to 0.5 inches.
4. Name the toolpath **wheel_thinner**.

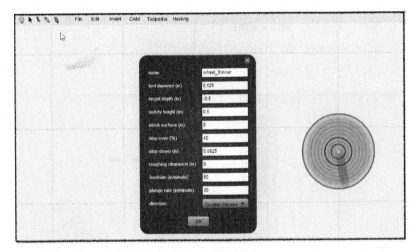

Figure 6-11. *Pocket operation to reduce thickness of material*

Repeat this process for each of the four wheels. Alternatively, you can draw the first wheel, and then copy and paste the first wheel to create the other three wheels (navigate to Edit → copy and then Edit → paste).

After the pocket has been applied to the wheel, you'll need to create toolpaths for the other wheel features.

Screw Head Countersink

The wheel's center circle is a shallow pocket that makes the screw's head flush with the outside wheel surface, the digital equivalent of manually countersinking a part.

When setting the depth for this feature, you'll need to consider the original thickness of the material. Although our screw head pocket has a 1/8-inch depth (0.125 inches), this pocket needs to be 5/8 inches deep (0.625 inches), as the 1/2 inch of material above the starting point of the pocket has already been removed. Figure 6-12 shows both the MakerCAM view and the sectional view of our part for reference. To create the pocket toolpath:

1. Select all four wheels.
2. Navigate to CAM → profile.
3. Name the toolpath **countersink**.
4. Set the target depth to −0.625 inches.
5. Click OK to generate the toolpath.

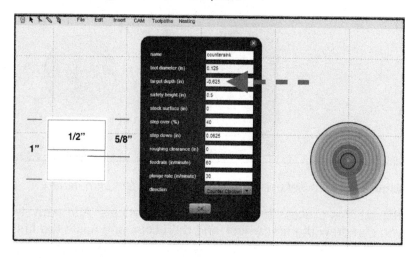

Figure 6-12. *Pocket operation to create recess for screw head*

Screw Hole

After defining the pocket for the screw, add the inside profile operation to create a hole for the screw to go through.

Step 5: Create Body Toolpaths

To create the toolpaths for the racer body, you'll use three operations, as discussed in the following sections.

Body Window

Navigate to CAM → inside profile. Use the screenshot in Figure 6-13 as a guide when setting your parameters.

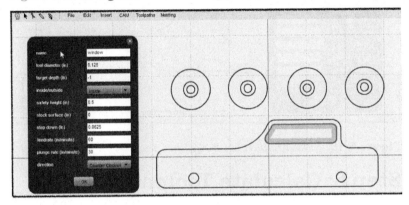

Figure 6-13. *Racer body window toolpath*

Wheel Holes

The holes for the wheels are also *inside profiles*. Use the screenshot in Figure 6-14 as a guide when setting your parameters.

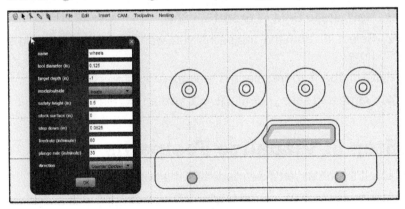

Figure 6-14. *Toolpath for racer wheels*

Body Perimeter

The perimeter of the body is an *outside profile*. Use the screenshot in Figure 6-15 as a guide when setting your parameters.

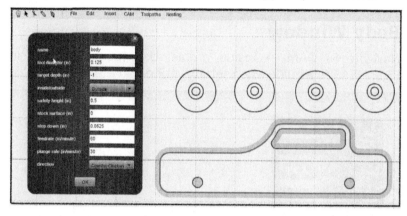

Figure 6-15. *Toolpath for racer body*

Step 6: Calculate Toolpaths

After all the toolpaths have been created and saved, you'll need to calculate them.

Navigate to CAM → calculate all.

Step 7: Export G-code

1. Navigate to CAM → export G-code.
2. Select or add the toolpaths you want to mill at the same time.
3. Click the "export selected toolpaths" button.
4. Name and save your file; it will have an NC file extension.

Step 8: Vizualize Toolpaths

Now it's time to see if we made any mistakes!

Open your G-code file in a text editor and copy the contents. Then head to the G-code simulator (*http://nraynaud.github.io/*

webgcode/), paste in the text from your G-code file, and press the Simulate button in the lower-left corner of the interface (Figure 6-16).

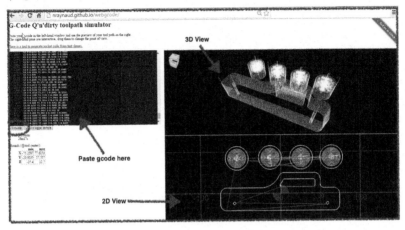

Figure 6-16. *Racer toolpaths in the G-code simulator*

The visualization panes on the righthand side of the G-code simulator are interactive—you can drag them around to change your view.

7/Creating Motion: Electromechanical Overview

Computer-controlled motion is the heart of a CNC machine. This chapter explains the mechanical components and machine design that make basic machine movement possible. You'll see how to create automated motion through the addition of motors, control boards, and machine-control software.

Mechanical Motion

This section explores the common types of linear components used in CNC machines, and how these linear guides driven by motors convert rotary motion into linear motion.

Linear Motion

One *linear rail* is required for each axis of a three-axis CNC. A linear rail is the generic term for a specialized bearing or slide designed to move with only one degree of freedom along a single axis.

There are many different types of linear rails, and although their appearance differs, their function is the same—to provide *slop-free motion*.

Slop

Slop is a linear rail moving in *more than one direction* because of bearing or rail wobble. Slop reduces accuracy and creates tremendous wear and tear on the machine, forcing parts to wear faster.

Your overall cut quality will also suffer. What should be crisp and perfect edges will look wavy, and the exact same depths entered into the CAM program may vary. In addition, your cutters aren't made to "give" or wobble, so you'll be much more likely to break bits.

A great example of slop can be illustrated by comparing drawer slides to a three-axis CNC. A drawer slide mechanism is a simple and effective design that works very well for low-precision applications such as toolboxes, silverware drawers, and filing cabinets.

Your drawer slides are optimized to slide smoothly forward and backward on one axis. However, if another force is introduced, like additional force from hastily pushing the drawer up or down when opening it, the drawer often gets "stuck" or "loose." Drawer slides that become worn from use or abuse wear out, and the components become looser, introducing a significant amount of slop.

This concept also applies to CNC machines. In order to move along two or three axes at a time, forces will be pushing against the linear rails in multiple directions.

Linear Guide Types

A few of the most common types of linear guides are shown in Figure 7-1.

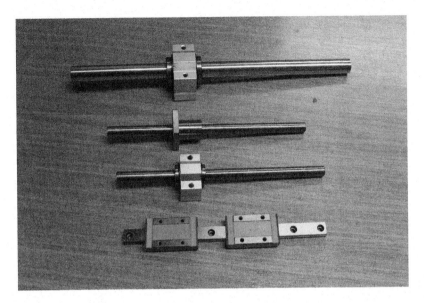

Figure 7-1. *An assortment of linear rails*

Now that the mechanical parts have been covered, the next step is to make those axes move automatically. While pushing and pulling the machine around by hand is an option, that's not what CNC is about. After all, it's supposed to be computer controlled!

Power Transmission

On any given CNC machine, each axis is individually connected in some fashion to a motor, which turns the shaft. This concept is called *power transmission*, and in this case it converts a motor's rotary motion in one axis to linear (side-to-side) motion in another axis, moving the machine.

Power transmission from rotational to linear motion in three-axis CNCs commonly occurs in one of three ways:

- Direct motor coupling with a lead screw
- A system of shafts, pulleys, and belts attached to a motor
- A rack-and-pinion configuration

Lead Screws and Lead Nuts

Lead screws and nuts have a different geometry than "regular" screws, bolts, and nuts. This geometry promotes smooth motion and does not bind like traditional screws and nuts.

Figure 7-2 shows a traditional lead screw and nut assembly. In this Nomad 883, the nut is fixed to the z-axis carriage, and the screw driving through the nut raises and lowers the carriage.

Lead nuts
   Internally threaded components that pair with the lead screw to complete a screw-driven system. Lead nuts are different from basic nuts, like the ones you find in a hardware store. Lead nuts are generally deeper than traditional nuts, providing more thread engagement. They are often made from brass or another self-lubricating material to reduce friction.

Lead screws
   By attaching a lead nut to a platform or toolhead, and then rotating a screw through the nut, this rotational motion will move whatever it's attached to (forward and backward, or up and down, in whatever direction the platform is intended to move). Different types of screws are commonly found in CNC machines, but they all have the same function.

Figure 7-2. *The Nomad CNC machine is driven by trapezoidal screws*

Belt Drives

Because of their low cost, high flexibility, and a wide variety of vendor options, belt-driven systems (Figure 7-3) have become very popular in hobbyist CNCs.

Figure 7-3. *Typical belt-driven setup*

At their core, belt-driven systems require at least four different components to function properly: a belt, a pulley mounted to a motor, and an idler. Somewhere in the system, the belt is connected to the carriage on one end and a pulley on the other.

When the stepper motor rotates, the teeth on the pulley engage with the teeth on the belt, moving the belt along a single axis.

Rack and Pinion

Rack-and-pinion systems work by mounting a *sprocket* (or gear-shaped device) onto a motor. The teeth of the sprocket fit into the corresponding teeth of the *rack*—a precision manufactured piece of metal with a tooth design on one side.

Rack-and-pinion systems can move very quickly and with a high degree of accuracy. They are more complex than belt or lead screw systems because they require additional components to properly tension and affix the components to the machine. They can be used in much longer linear applications than a screw. However, rack-and-pinion systems are rarely found on desktop-sized CNC machines.

Backlash

Think about riding a bicycle. Imagine you're pedaling along and decide to coast for a second. While you're coasting, you pedal backward, disengaging the chain-driven sprocket.

At a certain point, you decide to stop coasting and begin to pedal forward again, engaging the gears that drive the back wheel. As the pedal pushes the gears in a clockwise rotation, there will be a certain portion of the rotation of your pedals before you feel them re-engage—that distance is *backlash*.

In a screw-driven CNC machine, backlash is the distance required for the threads in the nut to re-engage with the threads on the screw when the direction of travel is reversed. In a belt-driven system, backlash would be the distance required for the pulley to re-engage with the belt after the direction of travel is reversed.

Eliminating all backlash in a system is impossible. However, most backlash can be removed by selecting the correct types of timing belts or thread profiles. Certain types of belting, such as MXL, GT2, and HTD, are made for linear motion, and are designed around the idea of decreasing backlash by optimizing the tooth profiles. For screw-driven setups, selecting an Acme (or other trapezoidal profile) along with an anti-backlash nut will reduce backlash immensely.

In addition to proper profiles and anti-backlash nuts, backlash can be compensated for in some software packages. Implementing backlash compensation depends on your software package, but generally the procedure involves using a dial indicator, setting zero, and measuring the actual distance moved after the direction has been reversed.

Motors and Electronic Components

A CNC machine needs a computer and the following components (see Figure 7-4):

Motors (specifically stepper motors)
   A machine, powered by electricity, that applies rotational movement by turning a shaft. Stepper motors are the most common type used in CNCs.

Stepper motor drivers
   A chip that separates power from the control board, to control how the motor turns in which direction. They can smooth out the motion of the stepper by providing fractional steps, keeping the motor from vibrating at certain speeds and potentially reducing resonance.

Power supply
   Supplies the voltage and current to power the stepper motors, stepper drivers, and machine control board.

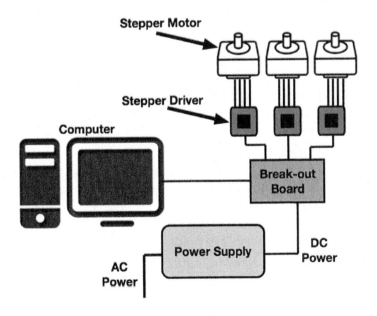

Figure 7-4. *CNC router—overview of a typical electrical system*

Powering and Driving Stepper Motors

It's important to note the relationship between stepper motors and power supplies. The correct power supply isn't determined by how many volts your stepper motor can handle; it's determined by how many volts your stepper driver can handle.

Most stepper motor drivers are current limiting, and known as *chopper* drivers. They're responsible for managing the current to the stepper and preventing too much current from ruining your motors. If you look at the label on your stepper motors, you'll find a voltage rating somewhere between 3 V and 5 V DC. However, if you look at the power supply, it's likely 24 V DC or higher. That extra voltage is managed by the stepper driver, as it's chopping the current.

Stepper Motors

In order to control a CNC machine, you must be able to control the motors that drive the axis. Controlling the motors entails both how far the motor turns and in what direction it is turning.

Industry has provided us with two types of motors made especially for this task: *servo* and *stepper motors*.

Servo motors are a special type of motor that have an encoder built in. The encoder is what gives a servo motor its magic. The encoder's job is to measure the number of rotations (or partial rotations) the motor has moved and compare that value to the distance the motor was expected to move.

This is known as a *closed-loop system*—the motor and the controller are able to stay synchronized with each other because of the feedback provided by the encoder. In a servo-driven system, the controller *always* knows exactly where the servo is, compared to where it's supposed to be.

For the hobbyist, or even prosumer markets, CNC machines are designed around stepper motors, and nearly all machines found in the sub-$5K range will use stepper motors because of their relatively low cost compared to servo motors, and the relative simplicity of the open-loop system.

An open-loop system means that a command is issued to the stepper motor, but unlike a servo, the stepper motor (because of its lack of encoder) cannot communicate back to the controller to update its position and be compared to the intended or expected position.

A stepper motor is a special type of motor that allows for control in both rotational direction (clockwise or counterclockwise) and distance rotated (number of rotations). The distance a stepper motor rotates is referred to as *steps*. Stepper motors sold for CNC machines are typically configured at 200 or 400 steps per rotation.

Why is it called a stepper motor? The nomenclature makes a lot more sense when you look inside a stepper motor. Once inside, it's easy to see why the name was chosen.

Rotation is produced by applying voltage in the correct order around the coils (see Figure 7-5), and controlling the direction of the current.

Figure 7-5. *Typical stepper motor shell*

Unlike a regular DC motor, which spins freely when voltage is applied, the stepper motor rotor (see Figure 7-6) moves one step at a time in a given direction.

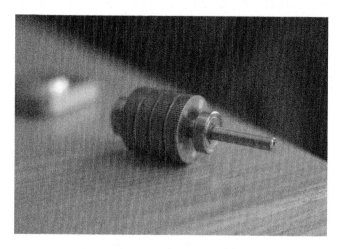

Figure 7-6. *Typical stepper motor rotor*

You'll find that most stepper motors are described as a *NEMA* value. The most common stepper motor found in 3D printers is the NEMA17. Some desktop CNC machines use larger NEMA23 stepper motors, while smaller 3D printers use NEMA14 stepper motors (Figure 7-7).

Figure 7-7. *Variety of stepper motor sizes*

This is the important part about NEMA values—they describe the *physical characteristics* of the motor's body. What a NEMA value does *not* tell you is anything about the motor's electromechanical properties.

Don't assume that motors with the same NEMA size are all equal. To know the exact specifications of your stepper motor, you need to find the vendor and part number and look at your motor's data sheet.

Motion and Machine Control

This is where the magic happens! Now that you know what the different parts of a CNC machine are, and what each one of them does for the overall system, there's only one thing left to figure out: how do all these components work together?

Looking at machine software broadly, two main things are happening:

- The *motion-control* software is planning your machine's moves and keeping everything synchronized.
- The *machine controller* is the software letting you control your machine.

In most cases, the same *machine software* package performs both tasks. Where the motion control is built into the machine controller, the two features appear to act as one.

It's important to know that two distinctly different tasks are being performed by the machine software—motion control and machine control. Let's briefly review each of these:

Motion control
　　Motion control is software that reads G-code and converts it into the electrical signals that tell the CNC to move. At its core, motion control is the link between software and hardware.

　　In most cases, users never interact with the motion-control software at all; it's hidden under the hood, and for most people, the hood is welded shut.

Machine control
　　This is the software that you interact with to control your CNC machine. Think of machine-control software as the cockpit for your machine!

Machine-control software lets you jog, home, load G-code files, set offsets, touch off, and everything else you need to do to operate your CNC machine effectively.

Because motion control is such a difficult problem to solve, only a few software solutions are available to hobbyist- and prosumer-level products.

The following sections list, in no particular order, the three most common machine software packages you'll find on today's desktop CNC machines.

Mach3

Mach3 (*http://www.machsupport.com/software/mach3/*) is a very popular Windows-based CNC controller that's been around since 2001. The newest version of the software is called Mach4, and although it has been released for sale, Mach3 (shown in Figure 7-8) is still the more popular of the two programs.

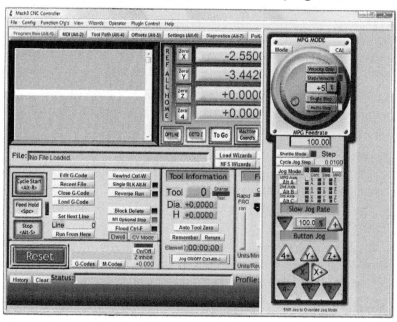

Figure 7-8. *Mach3 interface*

Mach3 runs only on Windows XP and Windows 7, and requires users to purchase a license.

It's important to note that Mach3 is also a machine controller, the interface to operate a CNC machine.

LinuxCNC

Just like Mach3, LinuxCNC (*http://linuxcnc.org/*) is a machine- and motion-control software package, but LinuxCNC (as the name indicates) is Linux based. The LinuxCNC interface is shown in Figure 7-9.

It's completely free and open and can be installed in a variety of ways, from a live-boot USB drive, to a fully installed image.

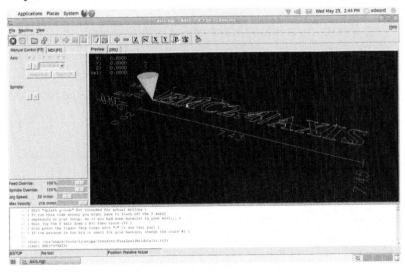

Figure 7-9. *LinuxCNC interface*

Grbl

Grbl (pronounced "gerbil," though you might hear other pronunciations, including "grr-ble" and "garble") is a completely different type of motion controller compared to Mach3 or LinuxCNC. Unlike the previous two software packages, Grbl does not require installation on a PC. In fact, it can be (and typically is) installed on a microcontroller platform like Arduino!

Unlike Mach3 and LinuxCNC, though, Grbl is *only* a motion controller. You can interact with Grbl via the terminal, but it requires another software package to control the CNC machine in the traditional sense.

In order to operate a CNC machine powered by Grbl (like Shapeoko), you need some sort of machine-control software. For Shapeoko, we wrote our own software called *Carbide Motion*, which allows Shapeoko users to interact with their machine by loading G-code files, jogging the machine, setting zeros, and running/pausing/resuming jobs.

Quite a few additional machine controllers also interact with Grbl:

- Will Winder's Universal G-Code Sender (*https://github.com/winder/Universal-G-Code-Sender*)
- Zapmaker's Grbl Controller 3.0 (*http://zapmaker.org/projects/grbl-controller-3-0/*)
- Vasilis Vlachoudis's bCNC (*https://github.com/vlachoudis/bCNC*)

Industrial Cases

Industrial motion control is a whole other animal. Industrial-class CNC machines do not use any of the motion-control solutions listed here.

All of the industrial machine vendors—for example, Haas, Morie Seiki, Mazak, and Fanuc—have proprietary motion-control systems they have developed (over decades) in-house and keep as well-guarded secrets.

Other Choices

Grbl is a relatively young G-code interpreter and is blazing the way for an entirely new market of CNC machines. Because of this, new machine-control software that works with Grbl is being developed every day.

Parallel Ports

Up until recently, in order to connect a CNC machine to a computer, you needed a parallel port. For anyone born after 1990, the parallel port is a 25-pin GPIO port that allowed ancient peripherals to be connected to ancient computers.

The parallel port offered stability and near instantaneous communication. Plus, with all of the port's available pins, there were plenty to control the stepper drivers, along with auxiliary commands like spindle, coolant, homing, E-stops, and other accessories like a relay to turn on dust collection.

Both Mach3 and LinuxCNC still rely on a parallel port to communicate with the controller. They do not support USB connections.

Grbl-based systems employ a different tactic. Instead of using the computer to parse the G-code and act as the motion controller, Grbl puts the motion controller on an Arduino! Instead of using the parallel port to connect to stepper drivers and relays, the computer simply connects to the Arduino via USB, and the Arduino connects to the stepper drivers.

Both methods have their advantages, but it's important to know that as you're looking around the Internet for CNC information, there is no doubt you will run across something saying you need a computer with a parallel port to operate your CNC machine. This is true only if you're using Mach3 or LinuxCNC.

8/G-Code: Speaking CNC

G-code is the generic name for a plain-text language that CNC machines can understand.

 G-code actually uses many letters, not just G. I'll explain them each in turn.

G-code began as a project developed at MIT in the late 1950s. Because of the project's organic growth, there is no *universal* G-code standard that everyone uses. There are many different standards, and not every machine understands each one. The *flavor* of G-code you use depends on the machine you're using.

Standards compliance is less of an issue when using G-code generated by most modern-day CAM packages, which offer multiple *postprocessors* to chose from. A postprocessor outputs G-code appropriate for specific machines or machine types.

 Using a modern-day desktop CNC machine and software, you'll never have to enter G-code manually, *unless you want to*. The CAD/CAM software and the machine controller will take care of all of this for you. However, some people (especially Makers!) like to know what's under the hood and how things really work.

To look at an example of G-code, you can open any of the files your CAM program exports with a plain-text editor (see Figure 8-1); alternatively, you can export G-code from Maker-CAM, as described in "Step 7: Export G-code" on page 92.

You'll see that a G-code file is nothing more than plain text. It's not exactly human readable, but if you learn enough about the commands, it's pretty easy to read through the file and figure out what's going on. I'll present some examples of G-code and explain the most common G-code commands.

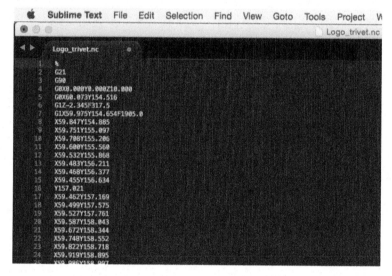

Figure 8-1. *G-code viewed from Sublime Text text editor*

Drawing a Square: Instructions for Humans

Imagine telling someone how to make a square with a pen on a piece of paper—it might go something like this:

1. Put pen down on paper
2. Move pen:
   a. Move pen 1 inch toward the *top of the page*
   b. Move pen 1 inch toward the *right edge of the page*
   c. Move pen 1 inch toward the *bottom of the page*

d. Move pen 1 inch toward the *left edge of the page*

3. Lift pen 1 inch from the paper

If these instructions are followed correctly, they'll find themselves with a nice square drawn on the page (see Figure 8-2).

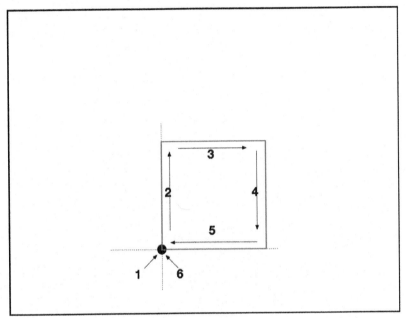

Figure 8-2. *Illustrated pseudocode commands*

But CNC machines don't understand plain English. In order to tell the machine what to do, you need to use a language it understands: G-code.

Square-Drawing Instructions for Machines

Let's use our pen example again, but this time, we'll translate from English to G-code (see Table 8-1).

Table 8-1. *G-code equivalents for drawing a square*

English	G-code
Put the pen down on the paper	G20 F20 X0 Y0 Z0
Move the pen 1 inch toward the top of the page	G1 Y1
Move the pen 1 inch toward the right edge of the page	G1 X1
Move the pen 1 inch toward the bottom of the page	G1 Y0
Move the pen 1 inch toward the left edge of the page	G1 X0
Lift the pen 1 inch from the paper	G1 Z1

Take a closer look at what the information in "Square-Drawing Instructions for Machines" on page 113 actually means. In the following sections, you'll look more closely at these instructions and describe the commands for each of the steps.

G-code Square Breakdown

Step 1: Put Pen to Paper (G20 F20 X0 Y0 Z0)

Let's break down the "put the pen down on the paper" instruction set—G20 F20 X0 Y0 Z0:

G20
   Inches versus millimeters. G20 tells the machine that you're using Imperial units (inches).

 If you were working in millimeters, you'd use G21.

F20
   F stands for feed. F20 tells the machine to move laterally 20 inches per minute.

Refer to "Speeds and Feeds" on page 49 if you need a refresher on feeds and speeds.

X0 Y0 Z0
> The x- and y-axes move to where the tool was zeroed (see "Zeroing the X- and Y-Axes" on page 128).
>
> Instructs the z-axis to move to the surface of the material.

Step 2: Move the Pen 1 Inch Toward the Top (G1 Y1)

Combining G1 and Y1 tells the machine to move 1 inch "away" from you (toward the top of the page) at 20 inches per minute:

G1
> Tells the machine to move at the feed rate specified in step 1 (F20), which in this context means 20 inches per minute.

 For reference, G0 tells the machine to move as fast as it can. G1 is a speed-controlled move.

Y1
> Move the toolhead to 1 inch in the positive *y* direction, *from the current toolhead's position*. You are currently at Y=0.
>
> For most machines (if you are standing in front of the machine), the *y* direction is front and back. A positive number moves the head farther away from you, and negative numbers move the machine closer to you.

Step 3: Move the Pen 1 Inch Right (G1 X1)

Similarly, G1 paired with X1 moves the pen 1 inch toward the right edge of the page (G1 X1):

G1

> As mentioned previously, G1 instructs the machine to move at the feed rate of 20 inches per minute, as specified in step 1 by F20.

X1

> Moves the machine 1 inch in the positive *x* direction—from the toolhead's current location (you are currently at X=0).

 From the front of most machines, the x-axis positive moves are to the right, negative moves to the left. G1 X1 tells the machine to move 1 inch to the right at 20 inches per minute.

Step 4: Move the Pen 1 Inch Toward bottom (G1 Y0)

The (G1 Y0) pair moves the pen 1 inch toward the bottom of the page:

G1

> Again, tells the machine to move at 20 inches per minute (because, again, that's what you told it to do in step 1).

Y0

> This tells the machine to go to Y0 in the *negative y* direction. This will move the head of the machine back toward you!

Most G-Code Moves Are Absolute

You may be thinking to yourself: "Why does Y0 make the machine move at all? I thought Y0 would move the machine zero units in the *y* direction!"

By default, moves in most G-code are *absolute*. Remember that machines move according to specified coordinates. G-codes specify location, not direction.

Y0 says to *move to* y-axis coordinate 0. If at this point in the exercise you told the machine G1 Y1, the machine wouldn't move! Because it's already at Y1.

If you told the machine to go to Y-1, the machine would move 2 full inches! (from 1 to -1). With the machine at Y1, if you tell it to go to Y0, it will move *from* 1 *to* 0.

> The coordinates are absolute—that is, they are not relative to the machine's last move.

Step 5: Move the Pen 1 Inch Left (G1 X0)

Move the pen 1 inch toward the left edge of the page (G1 X0):

G1
Tells the machine to move at 20 inches per minute (because you defined F20 earlier).

X0
Tells the machine to move to X0. Because the machine is currently at X=1, giving the X0 command will move the head of the machine in the negative *x* direction (to the left) 1 inch. This command will move the head of the machine in the negative *x* direction (or to your left) 1 inch.

Step 6: Lift the Pen 1 Inch from Paper (G1 Z1)

Used in combination, commands G1 Z1 pull the pen up from the paper, so that any subsequent movements will not draw:

G1
Tells the machine to move at 20 inches per minute.

Z1
Tells the machine to move the head to positive 1 inch.

In a nutshell, that is G-code! Telling the machine (very specifically) where to move and how fast to do it. If you are interested in learning more about G-code, a great exercise to do with any CNC machine is to manually send it G-code commands.

G-code Rules

Just like a math equation, G-code has its own rules about the order of operations. Here are the most common, in order of precedence (that is, comments will be interpreted first and and the change tool will be interpreted last):

Highest	Comments
	Feed rate
	Spindle speed
	Select tool
Lowest	Change tool

Feeds, Speeds, and Tools

The following are descriptions of the gcodes commands used for setting the speed, feed, and tool parameters.

"F" stands for "feed"
   The F command sets the feed rate; the machine operates at the set feed rate when G1 is used, and subsequent G1 commands will execute at the set F value.

   If the feed rate (F) is not set once before the first G1 call, either an error will occur or the machine will operate at its "default" feed rate. An example of a valid F command:

   G1 F1500 X100 Y100

"S" is for "spindle speed"
   The S command sets the spindle speed, typically in revolutions per minute (RPM). An example of a valid S command:

   S10000

"T" stands for "Tool"
   The T command is used in conjunction with M6 to specify the tool number to be used for the current file. An example of a valid tool change command:

   M6 T1

M6 T = Tool Changes

On industrial machines, an `M6 T` command usually produces a tool change with an automatic tool changer. On hobby machines with no tool changer available, issuing a new `M6 T` command will generally cause the machine to issue itself a feed-hold command, wait for the operator to change the tool, and then continue the job after the "resume" button is pressed.

Diving Further into G-Codes

G-codes tell the controller what sort of motion is desired. Some example motion types are *rapid*, *controlled*, *helical*, and *arcs*.

When you issue a G command, you are putting the machine into that *mode*. If you issue a G1 command, such as `G1 X5 Y13`, the machine moves to Y5 Y13.

If you issue another set of coordinates, you do not need to issue another G1 command. Why? Because the machine is in G1 mode until you change it to something like G0 or G2 or G3.

G0 (Rapid Motion)

The G0 command, as shown in Figure 8-3, moves the machine at max travel speed to whatever coordinates follow G0. The machine will move in a coordinated fashion, and both axes complete their travel at the exact same time.

G0 is *not* used for cutting. Instead, it's used to move the machine quickly to begin a job or move to another operation within the same job. Here is an example of a rapid (G0) command:

```
G0 X7 Y18
```

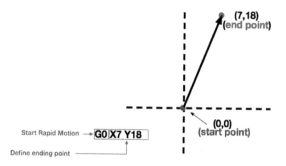

Figure 8-3. *G0 moves the machine at max travel speed*

G1 (Controlled Motion)

As illustrated in Figure 8-4, issuing a G1 command tells the machine to move at the specified feed rate (F):

G1 X7 Y18 F500

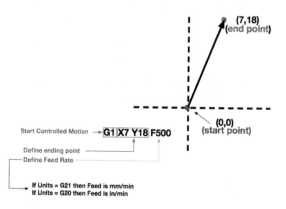

Figure 8-4. *G1 moves the machine at a specifed feed rate*

G2 (Clockwise Motion)

As shown in Figures 8-5 and 8-6, setting the mode to G2 and specifying the offset from the center point creates clockwise motion between the starting point and the specified ending points.

```
G21 G90 G17
G0 X0 Y12
G2 X12 Y0 I0 J-12
```

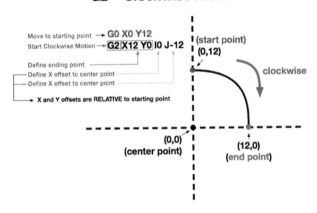

Figure 8-5. *G2 motion example*

The G2 starting point is where the machine is located prior to issuing the G2 command.

 It's easiest if you move your machine to the starting point *before* trying to issue to the G2 command.

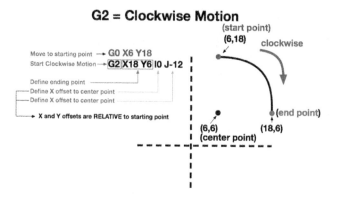

Figure 8-6. *G2 clockwise motions*

G3 (Counterclockwise Motion)

Just like G2, the G3 command creates an arc between two points. Whereas G2 specifies clockwise motion, G3 specifies counterclockwise motion between the points (Figure 8-7). A valid set of commands to produce G3 motion is shown here:

```
G21 G90 G17
G0 X-5 Y25
G3 X-25 Y5 I0 J-20
```

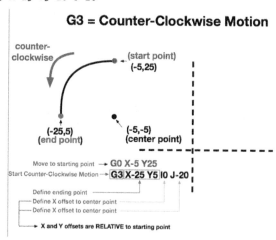

Figure 8-7. *G3 counterclockwise motion*

> ### Arcs
> With both G2 and G3, if the ending point is the same as the beginning point, a full 360-degree circle will be created. These two modes are not limited to *quarters*—any increment of degrees can be specified.

G17/G18/G19 (Working Planes)

These modes set the plane to be machined. Typically G17 is used and is the default for most hobby machines, but for reference here are the descriptions of the other two planes that can be used in a three-axis machine:

- G17 = x/y plane
- G18 = z/x plane
- G19 = y/z plane

G20/21 (Inches or Millimeters)

The G21 and G20 commands determine the G-code units, either inches or millimeters:

- G21 = millimeters
- G20 = inches

Here is an example of G-code set to millimeters (G21):

    G21 G17 G90

G28 and G28.1 (Referencing Home)

On some machines, issuing a G28 command will send the machine to its home position. Other machines may require a G28.1 command, along with the coordinates: G28 X0 Y0 Z0.

On Grbl-based machines, issuing a $H command from the MDI prompt will initiate the homing sequence.

G90 (Absolute Mode)

G90 causes units to be interpreted as *absolute coordinates*. This is the most common mode for hobby-grade CNC machines; it's the "default" mode.

Absolute coordinates will be interpreted as exactly what their name indicates—absolute. G0 X10 will send the machine to X=10. It will not send the x-axis to "10 more" units from where it's currently located.

G91 (Incremental Mode)

The opposite mode of G90. Setting *incremental mode* means that every command issued will move your machine the specified number of units from its current point.

For example, in incremental mode: G1 X1 will always advance the machine 1 unit in the *x* direction, regardless of its current location.

M-Codes

M-codes are action codes. Each M-code listed here tells the machine to physically do something, like turn an accessory on or off, or to stop completely:

M00
    M-Code Program Stop (nonoptional)

M01
    M-Code Optional Stop: Operator Selected to Enable

M02
    M-Code End of Program

M03
    M-Code Spindle On (CW Rotation)

M04
    M-Code Spindle On (CCW Rotation)

M05
    M-Code Spindle Stop

M06
    M-Code Tool Change

M07
    M-Code Mist Coolant On

M08
    M-Code Flood Coolant On

M09
    M-Code Coolant Off

9/Practical Machining Tips

Even with a firm understanding of the workflow and steps required to go from idea to finished product, you're going to run into some situations where you're left scratching your head, wondering how to create a part or overcome a CNC machining constraint. Becoming proficient at designing for and then actually executing your ideas isn't something you learn overnight.

CAM File Orientation Versus Actual Machine Setup

When you design in CAD software, you probably position your drawing relative to the CAM program's orientation and layout. However, when you import that file into a CAM program, you may need to adjust the placement of the vectors. For example, you may have created your drawing in Landscape view, but the actual machine is oriented in Portrait view.

Figure 9-1 shows five possible orientations for placing parts relative to the x and y origin, but in radically different places in MakerCAM.

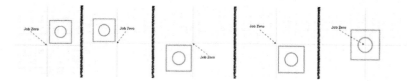

Figure 9-1. *Job zero examples in MakerCAM*

Understanding the relationship between the way files appear in a CAD or CAM program versus where the toolpaths will begin when you run the G-code is critical to a sucessful machining job.

Setting Machine Zero

For people new to CNC machining, it's a common mistake to forget to zero the machine before beginning a job.

You've loaded your G-code file and jogged your cutter to the bottom-left corner of the stock material. You think you're ready to go, so you hit the big green Go button.

You were anticipating your machine to begin cutting exactly where the spindle is located. But suddenly, your machine darts off into some "random" direction. You immediately hit the E-stop button, but are left scratching your head, trying to figure out what just happened!

In order for your G-code to begin cutting at the corresponding zero coordinate for both the file and the machine, you need to zero your CNC in all three axes. That is the point where the G-code file will start.

Zeroing the X- and Y-Axes

You need to determine the best place to set job zero for the x- and y-axes, and it's easiest if you always use the same place. The two most common zero locations are the center or the bottom-left corner of the cutting area:

Center of part
You move the spindle to the center of your material, and then use your machine-control software to set the x- and y-axes to 0.00. See Figure 9-2.

Figure 9-2. *Origin at material center*

Bottom-left corner of part (first pane)
    In this setup, you move your spindle to the lower-left portion of your stock material, and then use your machine-control software to set the x- and y-axis to 0.00. See Figure 9-3.

Figure 9-3. *Origin at bottom-left corner*

It's sometimes helpful to measure your material, marking the center with a pencil to aid bit alignment when setting x,y to 0.

A quick and easy way to do this with square or rectangular stock is to simply draw two lines in an X across the material. Your center zero will be the point where the lines intersect.

Zeroing the Z-Axis

Zeroing the z-axis is conceptually the same as zeroing the x- and y-axes, except there's a special time-tested procedure that will help you set your z-axis to zero accurately.

 The most common z-axis zero point is the *top* of the material. When the top of the material is the zero point, all of your cutting depths will be a negative value.

After positioning your x- and y-axes to the correct location, and then setting them both to 0.00, begin lowering your spindle (slowly) toward the workpiece.

As you get closer to the workpiece, place a piece of paper on top of your material and begin slowly moving it back and forth. Continue lowering the z-axis slowly until you can no longer move the paper.

At this point, your cutting tool is sitting exactly on top of the material, and you can set your z-axis to 0.00 from within the machine-control software.

Once your z-axis has been set to 0.00, use the control software to raise it slightly and remove the paper.

Now all three of your axes are zeroed!

Finding Edges

If you're machining a part that is already made (say, engraving a design on the back of your iPad), finding the exact edge of the part is very important.

A special tool called an *edge finder*, shown in Figure 9-4 (sometimes called a *wiggler*), is used for finding the exact edge of a part. Edge finders come in both mechanical and electronic versions and are fairly common, even in hobbyist-level CNCs.

A note about edge finders: most are intended to be run at or below 1,000 RPM. If you are using a CNC machine with a trim router, odds are you cannot get the trim router to spin that slowly. In this case, you'll need to either use an optical edge finder or manually touch off the edge.

Figure 9-4. *Basic edge finder with 1/4-inch shank*

Once the edge finder locates the edge, its diameter tells the operator the offset from the center of the spindle to the edge of the part.

Homing

Homing gives you a *physical* reference point on your machine that never changes. By moving your machine to the *home position*, you create a starting point for referencing other locations.

> ## Switches
>
> In the Maker lexicon, you will often hear the terms *limit switches* and *home switches* used interchangeably. Technically, this is incorrect, as home switches are the microswitches used to locate the home position, while limit switches are triggered when the machine reaches the physical travel limit of one of your axes.
>
> However, it is very common that these switches are either the same switch, *or* there is only one set of switches performing only one of the functions, typically homing.

Homing is activated differently depending on the machine-control software you are using. You'll need to look around the interface of your controller and find the "home" button to activate the homing sequence. Once you initiate the homing sequence, the following actions will more than likely take place:

1. Raise the z-axis until z-axis home/limit switch is triggered. This is the maximum positive travel for the z-axis.

2. Move the x- and y-axis in the direction of the home/limit switch.

3. When the switches are triggered, the axis that engages them will back off slightly (to release the switch) and then move very slowly toward the switch, engaging the switch a second time. After the second engagement, the carriage will back off a set distance (1–5 mm).

Practical Homing

It's rare that you would start your job from the machine's home position. Instead, the home position is used as a reference to move your machine to the location where you wish to start your job.

In real-world usage, you would generally home your machine, and then jog the machine to your job's starting location. Once the machine is positioned where you want to begin, you set your x- and y-axes to 0.00, and your z-axis to 0.00 by following the z-axis zeroing procedure outlined in "Zeroing the Z-Axis" on page 130.

This position you move *to* from the home position is the offset from the home position. If you were to write down those coordinates (the offset) and then re-home your machine, you could move back to the offset location by typing in the coordinates you wrote down!

Tool Changes

Imagine you have a part to make that requires two different-size bits:

- 1/4-inch square end mill to do a roughing pass
- 1/8-inch ball-nose end mill to do a finishing pass

Here's how to ensure that the toolpaths properly align:

1. Home your machine.
2. After homing, jog the spindle to your job's starting point (typically the bottom-left corner or the center of the material).
3. Make a note of the machine's current x,y position. Then set it to zero from within your machine control software.
4. Zero your z-axis.
5. Run the roughing pass job with the 1/4-inch end mill.
6. After the job completes, swap out the 1/4-inch cutter for the 1/8-inch cutter, and then home your machine again.
7. After the machine is homed, move your machine back to the exact coordinates you noted in step 3.

8. Set the x- and y-axes to zero in your machine-control software, and then re-zero your z-axis by following the procedure outlined earlier.

9. Then you would run the rest of your job with the 1/8-inch ball-nose end mill!

G-Code: Tool Changes

There are a couple of common ways tool changes are implemented. For most people, it's easiest to create toolpaths for each tool size and export a different NC (G-code) file for each tool.

The other option is to have your CAM program export all of your G-code in one file, but call a *toolchange* in G-code when it's time to change bits. Calling a toolchange is done by issuing the following G-code command:

    M6 T##

M6 means *stop everything and change the tool*.

T tells us the tool number (##). Tool numbers are commonly used on industrial machines that have the ability to automatically change a tool.

Machined Material Hold-Down Tips

Aside from methods of holding your material to your work table, it's also important to secure your part to the material. As your job progresses, the machine is actually cutting your part *away* from the stock material. If proper precautions are not taken, when the machine cuts all the way through the stock, there will be nothing left holding your part down! The stock material may still be held down to the table, with whatever methods you decided to use, but the part that you are making would not be secured. There are two pretty common ways to deal with this:

Pre-drilling and using screws
   This first method is to place screws inside the profile of your finished part. This may be difficult to do without interfering with your toolpath. But for complex parts or parts that need

a very clean finish on the edge without secondary finishing required, holding the part from within is a great option.

Holding tabs

Creating tabs is a built-in function to most CAM packages. Tabs keep a small part of the stock material connected to the finished part, as shown in Figure 9-5. By leaving the two connected, the finished part remains secured to the stock material during the entire cutting process and will not break free.

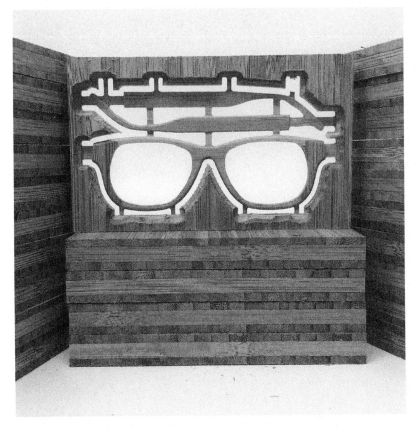

Figure 9-5. *Holding tabs keeping finished sign secured to stock material*

A secondary process of removing the tabs is required. Generally, removing the tabs can be done with a sharp chisel, flush-cut pliers, or even a Dremel with a small cutoff wheel (for materials like aluminum). Tabs are the simplest and easiest method of keeping your finished part secured to the material.

10/Conclusion

The world of CNC is enormous. Depending on what your goals are and where you're starting your journey, it may seem daunting or even impossible. This chapter provides some tips to help get you started.

I've sold thousands of CNC machines over the last four years. People of all skill levels have bought my machine and used it for amazing things. They have used Shapeoko for things that I never would have imagined it would be used for.

I've tried to talk to and interact with as many Shapeoko owners as I can. Over the last few years, there are some patterns that have surfaced that I'd like to share with you:

- The people who are most successful with their CNC machine are the ones who buy the machine for a purpose. They intend to make a specific thing, and in most cases have gotten to the point where Shapeoko is the final thing they need to complete their idea. This holds true for both users with and without previous experience. As long as they have a specific *project* in mind, they almost always succeed!
- Shapeoko owners who have experience with other types of CNC machines generally do well with Shapeoko. They seem to understand that a workflow is required to go from idea to finished product and can usually transition with very little friction from their experience using another machine to using Shapeoko.
- The people who are most unsuccessful are the ones who have zero experience and do not have a specific purpose for

the machine. These customers generally fail early in the process and have a difficult time getting up to speed.

After talking with many different types of customers, it's clear to me why the trend is true and why it continues: scope! Getting started with CNC is a huge undertaking! It's so enormous that it's hard to know where to begin and how wide of a net to throw when you're looking for information.

Customers who don't have experience and don't have a specific product tend to get lost in the noise and bogged down with details that in actuality don't apply to them. With no experience, it's hard for them to discern what is important information and what is not. Frustration sets in quickly, and the task seems too hard to manage.

Customers who have experience understand what they need to "figure out" in order to make their next step toward being proficient with the machine. They can use their previous experience as a reference point for learning something new.

For customers with zero experience but who have a specific project, they are self-limiting their scope, creating their own constraints. All they need to do is figure out how to make the one thing they bought the machine for. Once they learn that, it's a fairly incremental task to add something to their design and then learn how to translate that to CNC.

There are thousands of other topics related to subtractive CNC that never made it into this book. I tried to trim this down to the bare essentials of what you need to know in order to get up and running, making whatever you wanted, on your own CNC machine. I think it's safe to say that if the topic is not in the book, it shouldn't be a roadblock for you to get started, because it's probably not required.

Whatever you're doing with CNC, I hope this book helps get you started.

A/Resources

On several occasions since releasing Shapeoko in 2011, I've had the opportunity to answer the question, "Where did you learn this stuff?" To which, I love to answer: "Everywhere!"

The phrase "Standing on the shoulders of giants" has never been as relevant to anything in my life as it is to CNC. From my father and his roots in manufacturing and machining to all of the wonderful people I've met along the way who have been so kind as to share with me their knowledge, which helped me learn that one thing that was crucial to learning the next "one thing."

Shapeoko Documentation and Communities

The following documentation and community information will prove useful as you get started with Shapeoko:

Shapeoko GitHub (https://github.com/shapeoko)
   Access to the raw BOM, build instructions, documents, drawings, assemblies, and datasheets for the Shapeoko and Shapeoko 2.

Shapeoko wiki (http://www.shapeoko.com/wiki/index.php/Main_Page)
   The wiki includes step-by-step build instructions and upgrade possibilities for numerous machine types. The wiki for the first machine is available at the Assembly Overview (S01) page (http://www.shapeoko.com/wiki/index.php/Assembly_overview_(SO1)).

Shapeoko forum (http://www.shapeoko.com/forum/)
> The community that started around the original Shapeoko project has since grown to thousands of active members. The Shapeoko forum is known for its welcoming atmosphere, technical excellence, and a collection of users who are both starting out and have been in the game for a long time. It's one of the few CNC places on the Internet that doesn't look down on newcomers, and to the contrary, welcomes them. Some key members of the community are Catalin Voinescu, Brandon Fischer, Tim Foreman, Will Adams, and Winston Moy. Each has thousands of posts, has created amazing projects, and is always available to help a new user get started.

Invaluable Resources

I found the following resources extremely useful during my research for this book and during my exploration into CNC:

Guerrilla Guide to CNC Machining, Mold Making, and Resin Casting (http://lcamtuf.coredump.cx/gcnc/)
> This guide, written by Michal Zalewski, is a gem of information. With plastic molding as the end goal, Michal explores the entire system of CNC—from machine selection to software options to mastering CNC design principles for both general purpose and mold making specifically. Even if mold making isn't your finish line, this website is full of very relevant and very useful information.

BuildLog.Net (http://buildlog.net/)
> Bart Dring's professional expertise as an engineer and his love and passion for making things sparked an unbelievable community of Makers. Both on the blog, where Bart has posted amazing projects that most Makers could only dream of building, and in the forum, information can be found that will help you along on your own CNC journey. 3D printers, laser cutters, CNC mills, and anything else that speaks G-code are all fair game at BuildLog.Net.

MakerSpaces

All across the country, and all around the world, MakerSpaces have been and still are popping up. If you're interested in CNC, chances are someone at your local MakerSpace is too. From there, you can either learn as a group, or maybe you'll be lucky enough to find a resident expert. Just remember—if they have a certification class, be sure to enroll before you try to fire up the machine!

Index

Symbols

$H command, 124
100kgarages, 15

A

absolute coordinates, 124
accuracy, 2
acknowledgments, xiv
Adobe Illustrator, 59
arcs, 123
Arduino, 109
Aspire, 68
AtFab, 14
AutoCAD, 59
Autodesk Fusion 360, 62, 68
Autodesk Inventor, 62
automatic tool changes, 119, 133
available travel, 24
axes, 6

B

backlash, 100
ball-nose end mills, 43
bCNC, 108
Bell, Cara, 10
Bell, Stephen, 10
belt drives, 99
bitmaps, 56
bits (see end mills)
Blender, 62
blocks, 10
board games, 10
body (spindle), 24
Bridgeport mills, 27
Buildlog.net, 140

C

CAD (computer-aided design)
    2D raster images, 56
    2D vector graphics, 57-59
    2D vs. 3D, 60-61
    3D models, 62
    halftone images, 65
    image to G-code, 63
    role in digital fabrication, 2
    single line drawing, 64
    tutorial, 81-93
    V-carving, 63
CAM (computer-aided manfacturing)
    basic operations, 79
    CAM file vs. machine orientation, 127
    defining toolpaths with, 67-73
    minimum feature size, 77-79
    role in digital fabrication, 2
    toolpath simulation in, 3
    tutorial, 81-93
CamBam, 67
Camillette Contemporary Jewelry, 18
CAMotics, 83
Carbide 3D, xv
Carbide Create, 67
carbide end mills, 46
Carbide Motion, 67
Carbide3D, 10
carriages, 24
Cartesian coordinate system, 6
carvings, 12
casts, 16
chipload, 50
chucks, 51
circles, 123

circuit boards, 20
clamps, 29
clearance height, 73
climb cuts, 49, 73
closed-loop motor control, 102
CNC (computerized numerical control)
   advantages of, 2-4
   approach to learning, xi
   prerequisites to learning, x, 138
CNC machines
   accuracy of, 2
   backlash, 100
   Cartesian coordinate system, 6
   defined, 1
   design capabilities of, 3
   homing, 132
   key parts, 23-28
   linear components in, 95-97
   locating machines near you, 15, 141
   machine configurations, 36
   machine vs. CAM file orientation, 127
   motion and machine control, 105-109
   motors and electronic components, 101-105
   possible uses for, 9-20
   power transmission in, 97-100
   safety guidelines for use, 4-5
   securing materials in, 28-36
   simulations with, 3, 77, 92
   tool changes, 133
   versatility of, 9
   X, Y, and Z axes for, 7
   zeroing, 128-131
CNC mills, 35
CNC routers, 36
   (see also routers)
CNC'ed Makerspace Shed, 15
code examples, using, xii
Coffland, Joseph, 83
collet (spindle), 25, 52-54
collet nut (spindle), 25, 53
comments, xiii
complexity, 3

contact information, xiii
contour finishing, 80
coordinates, 6
CorelDraw, 59
countersinking, 89
crossbow pistol, 11
cut depth, 72
Cut3D, 68
cutters (see end mills)
cutting techniques
   center cutting, 47
   chipload and, 50
   climb vs. conventional, 48, 73
   profile and pocket cuts, 69, 70
   ramping, 48
   speeds and feeds, 49

D

depth per pass, 50, 72
Desktop vacuum table, 34
digital fabrication
   defined, 2
   locating machines near you, 15, 141
   typical workflow, 2
dmap2gcode, 64
DOC (depth of cut), 50
dog-bone overcuts, 75
DolphinCAM, 68
DraftSight, 59
drawing (see CAD)
drill bits, vs. end mills, 40
Dring, Bart, 140

E

E-stops (emergency stop buttons), 4
edge finders, 131
electrical systems, 101
end mills
   alternate names for, 42
   center-cutting, 40
   coatings, 47
   common tool geometries, 42-43
   composition of, 46
   defined, 24, 39

vs. drill bits, 40
offsetting, 70
parts of, 45
recommended starter set, 47
selecting, 78
tip shapes, 45
tool changes, 133
types of tool holding, 51-54
engraving cuts, 70
engraving end mills, 43
ER collets, 52

F

F command (G-code), 118
F-engrave, 63
fabhub, 15
fair use, xii
feed rate, 50, 73
fill, 57
finishing
 contour, 80
 parallel, 79
fixed-diameter collets, 52
flat-tip end mills, 43
Flatworks LLC, 14
fly cutters, 43
Foreman, Tim, 11
formulas
 chipload, 51
 feed rate, 51
 speed, 51
Fortosis, Nicholas, 3
FreeCAD, 62
furniture, 14

G

G-code
 absolute, 116
 drawing a square with, 112-117
 exporting, 92
 F command, 118
 G0 (rapid motion), 119
 G1 (controlled motion), 120
 G17/G18/G19 (working planes), 123
 G2 (clockwise motion), 121

 G20/21 (inches/millimeters), 123
 G28/28.1 (referencing home), 123
 G3 (counterclockwise motion), 122
 G90 (absolute coordinates), 124
 G91 (incremental mode), 124
 history of, 111
 M (action) codes, 124
 M6 T command, 119
 motion types available, 119
 postprocessors for, 111
 rules for, 118
 S command, 118
 simulator for, 92
 T command, 118
games, 9
gantries, 24, 37
getting help, xiii, 139
Grbl, 107
GRBL Controller 3.0, 108
"Guerrilla Guide to CNC Machining" (Zalewski), 140

H

halftone images, 65
high-speed steel (HSS), 46
holding tabs, 135
home switches, 132
homing, 132
houses, 14

I

iDraw, 59
image2gcode, 64
incremental mode, 124
Inkscape SVG editor, 59, 61, 81, 85
inlays, 17-19
inside corners, 74
inside profiles, 70, 91

J

jaw chucks, 51
jewelry, 18

K

kerf, compensating for, 70
kill switches, 4

L

Lafreniere, Darren, 10
lead screws/nuts, 98
libreCAD, 59
limit, 24
limit switches, 132
linear rails, 95-97
LinuxCNC, 107
LittleMachineShop 3501, 37
Liyanage, Marc, 20

M

M (action) codes, 124
M6 T command (G-code), 119
Mach3, 106
machine control software, 105
machining tips
    CAM file vs. machine orientation, 127
    homing, 132
    patterns for success, 137
    securing machined material, 134
    setting machine zero, 128-131
    tool changes, 133
MakerCAM, 67, 81
MakerSpaces, 141
MeshCAM, 68, 79
metal creations, 17-19
mini-marshmallow pistol, 11
minimum feature size, 77
modeling (see CAD)
molds, 16
motion control software, 105
motors
    overview of, 101
    power supplies for, 101
    stepper motors, 102-105
moving carriage configuration, 36
moving gantry configuration, 37
moving table configuration, 36

N

neck (spindle), 25
NEMA (National Electrical Manufacturers Association), 104
nodes, 57
Nomad 883 CNC, 36, 99

O

Onshape, 62
open-loop motor control, 103
opendesk, 15
OpenSCAD, 62
operation parameter, 72
outside profiles, 70, 92
overcuts, 73-77

P

parallel finishing, 79
parallel ports, 109
parameters, toolpath, 71-73
paths, 57
peck drilling, 71
Pegs and Jokers board game, 10
permission, obtaining, xii
pinewood derby car, 10
pistol, mini-marshmallow, 11
pixelation, 56
platens, 28
plung rate, 73
PlyFly Go-Kart, 14
pocket cuts, 69, 70
postprocessors, 111
power supplies, 101
power transmission
    approaches to, 97
    belt drives, 99
    lead screws/nuts, 98
    rack-and-pinion systems, 99
printed circuit boards (PCB), 20
profile cuts, 69
profiles, outside vs. inside, 70, 91
Project Shapeoko, ix, xv, 36, 137, 139
projects
    circuit boards, 20

metal creations and inlays, 17-19
molds and casts, 16
signs and carvings, 12
toys and games, 9-11
vehicles, furniture, and houses, 14
wooden toy racer, 81-93
puzzle blocks, 10
PyCAM, 68

Q

qCAD, 59
questions, xiii

R

rack-and-pinion systems, 99
ramping, 48
raster images, 56
Raynaud, Nicolas, 82
resin casting, 16
resources, 139-141
Rhino, 62
routers
  vs. mills, 35
  speed control of, 26
  vs. spindles, 26
runout, 26

S

S command (G-code), 118
sacrificial layer, 28
safe z, 73
safety guidelines, 4-5
safety height, 73
Scorch Works, 63
screws, 32
servo motors, 102
shank (spindle), 25
shank end (spindle), 25
Shapeoko, ix, xv, 36, 137, 139
ShopBot Desktop vacuum table, 34
signs, 12
simulation, 3, 77, 92
single line drawing, 64
SketchUp, 62

slop, 96
software
  2D CAM packages, 67
  3D modeling, 62
  CAMotics, 83
  for desktop CNC machines, 106-108
  halftone images, 65
  image to G-code, 63
  industrial applications, 108
  Inkscape SVG editor, 81
  machine vs. motion control, 105
  MakerCAM, 81
  single line drawing, 64
  V-carve engraving, 63
  vector editing, 58
  Webgcode, 82
SolidWorks, 62
speeds and feeds, 49-51
spindle speed, 50
spindles, 24-27, 49, 52
spiral upcut/downcut end mills, 42
spoilboards, 28
square end mills, 43
step clamps, 29
step down parameter, 72
stepover, 73
stepper motors
  NEMA values and, 104
  powering, 102
  vs. servo motors, 102
  typical configuration of, 103
straight flute end mills, 42
strokes, 57
SVG (scalable vector graphics), editing software for, 59

T

T command (G-code), 118
T-bone overcuts, 76
T-slots, 31
table surfacing end mills, 43
tables (CNC work zones), 28
tables, vacuum, 33
tape, 33
target depth, 72

Ternus, Joe, 17
threaded inserts, 32
three-dimensional space, 6
TOhBaby.com, 10
tool changes, 133
tool diameter, 72
tooling (see end mills)
toolpaths
    2D/2.5D parameters, 71-73
    2D/2.5D toolpaths, 68
    3D, 69
    CAM operations for, 69
    creating, 67
    defined, 6
    overcuts, 73-77
    parallel vs. contour finishing, 79
    pre-machining simulations, 4, 77, 92
    tutorial on, 88-93
Tormach PCNC 1100, 31, 35
toys, 9, 81-93
tutorial
    creating digital design, 85
    MakerCAM configuration, 87
    project materials and dimensions, 84
    project overview, 83
    racer SVG file import, 87
    software needed, 81-83
    toolpaths, calculating, 92
    toolpaths, for body, 91-92
    toolpaths, for wheels, 88-90
    toolpaths, G-code export, 92
    toolpaths, visualizing, 92
two-dimensional space, 6
2D raster images, 56
2D vector graphics, 57-59
typographical conventions, xi

U

Universal G-Code Sender, 108

USB connections, 109

V

v-bit end mills, 43
V-carving, 63
vacuum tables, 33
variable frequency drive (VFD), 26
VCarve Desktop/VCarve Pro, 68
vector graphics, 57-59
Vectric, 67
vehicles, 14
vises, 34
Vlachoudis, Vasilis, 108

W

wasteboards, 28
Webgcode, 82
wigglers, 131
Winder, Will, 108
wooden blocks, 10
wooden toys, 9, 81-93

X

x-axis, 6-9

Y

y-axis, 6-9

Z

z clearance, 27
z travel, 27
z-axis, 6-9
Zalewski, Michal, 16, 140
zero, setting machine to, 128-131

About the Author

Edward Ford is a Maker. He was the kid who took his parents' TV apart to see how it worked, along with everything else in the house. Edward is trying to make the world a better place by empowering people to create their own high-quality products through advancements in desktop manufacturing.

In 2011, Edward designed and released Project Shapeoko via Kickstarter. Shapeoko has become one of the most popular desktop CNC machines on the market and continues to be a market leader and trendsetter.

In 2014, Edward helped cofound Carbide 3D, a company that specializes in the design and production of desktop manufacturing equipment such as the Shapeoko and Nomad lines of CNC machines. At Carbide 3D, Edward leads the Shapeoko product line and develops other great desktop manufacturing software and equipment.

Edward holds a B.A. in History from Rockford College.

Colophon

The cover and body font is BentonSans, the heading font is Serifa, and the code font is Bitstream's Vera Sans Mono.

CPSIA information can be obtained
at www.ICGtesting.com
Printed in the USA
BVOW11s0157160816
459147BV00001B/1/P